C0-AWL-520
FROM LIBRARY
WITHDRAWN FROM
TSC LIBRARY

A Chronology of Geological Thinking from Antiquity to 1899

by

SUSAN J. THOMPSON

The Scarecrow Press, Inc.
Metuchen, N.J., & London
1988

Library of Congress Cataloging-in-Publication Data

Thompson, Susan J.
A chronology of geological thinking from antiquity to 1899 / by Susan J. Thompson.
p. cm.
Bibliography: p.
Includes index.
ISBN 0-8108-2121-4
1. Geology--Chronology. I. Title.
QE11.T47 1988
550'.9--dc19 88-1493

Manufactured in the United States of America

CONTENTS

It is characteristic of science that the full explanations are often seized in their essence by the percipient scientist long in advance of any possible proof.

John D. Bernal

The greater and more revolutionary an idea, the more does it encounter resistance at its inception.

Pierre Teilhard de Chardin

In science the credit goes to the man who convinces the world, not to the man to whom the idea first occurs.

Sir William Osler

FOREWORD

This book began to take form during undergraduate Geology courses. Each course contained a brief historical overview but no one ever tied the bits and pieces together. I began merging my course notes at the end of each semester and found that the apparently disparate pieces did, in fact, suggest a coherent whole ... of innovative thinkers who were not always listened to, of simultaneous original and identical discoveries, and of persons lauded for refining or reviving someone else´s conclusions in a more favorable time. There are also many gaps which aroused my curiosity.

I decided to read more about the historical development of geological thinking but found references were few and scattered and difficult to identify since many bibliographic citations were absent, incomplete, and/or elusive to locate. I mentioned this to colleagues and they agreed and voiced a need for a bibliography tied to events as well as people, and small enough for student use but large enough for reference work. I have attempted to meet both needs.

Since my objective has been to record a maximum number of accomplishments and people, I used as many references as I could, regardless of whether they encompassed a sentence or several pages, expecting the reader to consult as many as could be located. (I dropped single-sentence references when I had several others to use.)

The arrangement is intentionally unorthodox. Entries are in chronological order to show the ebb and flow of discovery (and rediscovery). The date used is the date cited by the source except in instances where I knew that a reprint or collected volume date had been used. There I used the original date. Since the cited date is often the date a paper was presented, the journal citation date may be a year or two later.

I have entered the citations as they were given by the source of the quotation even when the title is translated or abbreviated, whenever I felt that the title is readily identifiable. (Pott´s *Lithogeognosia*, instead of *Chymische Untersuchungen Welche Furnehmlich von der Lithogeognosia*, for example.) Wherever possible, I have completed journal titles, pagination and publishing information. I did not attempt to verify all the citations. Only those which seemed suspect or on which two sources did not agree were checked. Where I could not identify the source of the quotation, I did not include it. The exception is manuscript, or early printed materials, which I included even when the title was unknown or the original lost. For the very old materials, I have not always included bibliographic information since any of a number of collected works would do. Whenever possible, I have included life dates, or century. Since more than one person sometimes independently discovered, or rediscovered, an idea, I have included all claims to original thinking. Similarly, I have not arbitrated when sources have interpreted a paper differently. I encourage the historians to determine who said what for the first time -- and what he really meant by it. Hopefully, I have perpetuated no errors or, worse, created new ones.

I have not quoted the source but have paraphrased his/her comments sufficiently to guide the reader to pertinent material but succinctly enough to keep the size of the Chronology manageable. These entries are simply intended to point the way to comments concerning the primary document in relation to a specific subject and person listed. When one source went into more or less detail than did another, I listed them separately rather than lumping them all under a single subject that would fail to identify ramifications mentioned by some but not by others.

Style of entry may vary somewhat but all entries should be sufficient to locate the material. I have followed the style of whatever bibliographic tool I used to verify the entry. For example, most journal titles were verified in the Union List of Serials but some were found in the World List of Scientific Periodicals, the International Catalogue of Scientific Literature (Section H, Geology), or in Scudder's Catalogue of Scientific Serials of All Countries Including the Transactions of Learned Societies in the Natural and Physical Sciences, 1633-1876. Books were verified in the National Union Catalog whenever possible but I also had to refer to the OCLC and RLIN databases, the Short-title Catalogues (Italian, German, French, English) of the British Museum, and the various national bibliographies of the countries concerned. Journal article titles were found in the Royal Society Catalogue of Scientific Papers, in Geologic Literature on North America 1785-1918 (U.S.G.S. Bulletin 746) as well as in the National Union Catalog in a surprising number of instances. Obscure titles may only be found in collections and library catalogs specific to the subject; for example, Geology Emerging (Ward and Carozzi, 1984), the catalog of the History of Geology (1500-1850) Collection in the University of Illinois Library.

I have included arguments for, but not against, a theory, unless the negative expression is itself a new theory, or the person expresses the opposite position elsewhere in the Chronology. I have excluded the patently ridiculous and irrelevant (such as the magical powers of stones) as well as folklore, lapidary, metallurgy and most applied sciences, such as industrial glasses and their impact on petrology. I have also excluded theological arguments, per se, even though the Church had a great effect on scientific development, or the hiatus therein. Also excluded are strictly descriptive works, blends of theories (unless the result was of greater importance than its parts) and local work unless it affected thinking on a vaster scale (for example, glacial work in Northern England and its effect on Agassiz' giant icecap theory).

Lastly, I have been unorthodox in the entry of names. Many people, especially students, do not know the rules of entry for De, La, Von, Van, etc., which may vary by nationality, historical period, and the work in which the name is cited. Therefore, I have entered all names by the smallest logical stem: Metherie, for example, instead of La Metherie, or De La Metherie, all three of which I have encountered during my research. Remember to try all possible arrangements when seeking additional information on these and other historical people. Also, anticipate variations in spelling of old and translated or transliterated names. I have used the commonest, or first encountered, here.

I hope this Chronology will help people to weave their own collection of historical bits and pieces into the coherent flow of discovery and rediscovery that is the recurrent theme of all history, to demonstrate that research is an organic process, ever changing and adjusting, and to integrate what we know today with why we know it.

Missing is that other great theme of history, the effect of social and religious values and beliefs on creative thinking. This is, in part, a reflection of work yet to be done. This book should be used in conjunction with works on social development so that our predecessors will be judged in the light of the handicaps under which they worked. I apologize in advance for those other omissions -- the events I did not read about or could not verify sufficiently to include. For every source I used, there were three more I could not use. For every citation I used, there were two more I could not use. I hope none were your favorites.

This book was facilitated by a Faculty Research Grant from Indiana State University, the Library Faculty Release Time for Research program, the determined (and often creative) efforts of Interlibrary Loan staff at Indiana University Northwest (Betty Shaw), Indiana University (Nancy Vossmeyer and Alice Crippen) and Indiana State University (May Ann Phillips, Karen Stabler, Lavedia Stout, and Judy Tribble). Thanks also to Sarah Gale for the bibliographic sleuthing and translating, and to many others too numerous to mention.

Susan Thompson
January 1988
Science Library
Indiana State University

1000/2000 BC

Vēdas (Knowledge), in Rig-Veda	Said the earth is a globe VAID 108
Vēdas (Knowledge), in Yajur-Veda	Said the earth circled the sun VAID 85

Twelfth Century BC

MANSUR, MOHAMMED BEN

Book of Precious Stones	Classified stones according to hardness and specific gravity BROME 120

Seventh Century BC

KANADA (Seventh Century BC)

Vaisēshika	Said there were elements which could never be destroyed, and which combine to form all that exists MUR 51

Sixth/Seventh Century BC

THALES OF MILETUS (624?-545 BC)	Said all things originated from water and even earth had once been water AD 9, 11, 124
	Said wind blew water into the earth from the oceans and then rock pressure drove it out as springs BAKER 396

Sixth Century BC

ANAXIMANDER (610?-547? BC)	Said all land creatures had developed from amphibians except for man who had developed from fishes AD 13
	Said fossil fishes formed from earth and heated water and man had formed from those same fishes HMPD 325
	Said time was endless but the universe is repeatedly destroyed and reborn DEAN 436

ANAXIMENES (545?-528? BC)

Said all things originated in air
AD 9

Said earthquakes were the result of earth drying out, splitting and uplifting mountains, after which water fills it again, causing the crust to swell
GEI 14

Said earthquakes were the result of large rocks within the earth tumbling and striking each other
AD 400

HERACLITUS (540?-480? BC)

On Nature

Said all things had their origin in fire but were constantly being destroyed and reformed into something else
AD 9; ZIT 4

PHILOPONUS, JOHN (Fl. Sixth Century BC)

On the Eternity of the World. Venetis?; N.P.

Said valleys and ocean basins were created to absorb the Noachian flood waters
AD 332

Fifth/Sixth Century BC

XANTHOS OF SARDIS (Fl. 500 BC)

Said fossil sea shells found inland proved the sea had once covered the land and that the land and sea were constantly changing places
AD 12; GEI 33; STN 73; ZIT 3

XENOPHANES OF COLOPHON (560?-478? BC)

Said fossils were once living and were proof that the land had been under water
AD 12; HMPD 324; P&S 64; RUD 39; STN 73

Said sedimentary rocks formed from debris the ocean had removed from the land
P&S 64

Said the ocean ate away the land
AD 12

Fifth Century BC

ANAXAGORAS (500-428 BC)

On Nature

Said metals grew in the crust of the earth
AD 289

Said a circular motion developed in the original chaos causing the universe to form differentiating ether, air, and water
AD 289

Said earthquakes were caused by ether sinking into the earth, encountering interior vapors and producing lightning and fire the latter of which forces its way surfaceward
AD 400; GEI 13

Said subterranean lakes fed springs and rivers
AD 426-7

ARCHELAUS (d. 428 BC)

Said air finds its way through passages into the earth´s interior where it is forced into underground caverns, compressing the older air, causing violent storms that move the earth and if air escapes to the surface, causes earthquakes which are especially common along the coast because water contains the air and it thrashes about seeking more room
AD 400-1

EMPEDOCLES OF AGRIGENTUM (490-430 BC)

On Nature

Said core of earth was molten
ZIT 5

Said earth developed in four progressive stages, plants before animals
AD 14

Said everything formed from the universal elements of the earth, water, air, and fire, and that everything is constantly changing
AD 13

Said huge fossil bones were those of now extinct human giants
HMPD 324

LEUCIPPUS (Fl. 400 BC)

The Great World System

Said all matter consisted of tiny particles, of varying size, shape, and mass, which are eternal and never still
MHG 196

PHILOLAUS (Fl. 450/475 BC)

(Fragments only remain of his work)

Said the universe consisted of five planets the sun, moon, and earth, plus a counter-earth and a central fire which were always on the back, or uninhabited side of the earth, and, far beyond, the stars
ZIT 5

PYTHAGORAS (580-500 BC)

(Quoted in) Ovid's Metamorphoses, Book 15

Said winds trapped in the earth lifted a flat plain up to form a hill
AD 330-331

Said earth was the center of the universal sphere and everything circled around it
AD 11; ZIT 5

Said earth's center was a fire which would eventually exhaust its fuel and die out
AD 11; GEI 37-9

Said nothing ever perishes, only changes form as sea and land exchange locations, mountains wear down to plains, lakes become deserts, etc.
GEI 37-9

Said running water cut the valleys
WOOD 4

Said world consists of four elements: air and fire above, water and earth below
GEI 37-9

450 BC

HERODOTUS (484-426 BC)

The History

Said black, crumbly alluvial soil of Egypt was deposited by a great river cutting through the sandy soil, stone and clay of nearby countries
HARR 457

Said shells found inland had been left there when the ocean covered the land
HMPD 324

Said that lower Egypt had been a gulf and that it had taken thousands of years to build up the delta which was still growing, as was the case in Greece and Turkey
AD 11; P&S 65; PRG 594

Said the one-celled fossils in some limestone were petrified lentils
FENT 19

Fourth/Fifth Century BC

DEMOCRITUS OF ABDERA (460-370 BC)

Said all matter is composed of atoms of varying size, shape, and mass which are eternal and constantly in motion, and change is the reorganization of atoms in accordance with rules of nature, since nothing can be destroyed
MHG 196; ZIT 56

Said heavy rainfall entering the earth´s cavity forced out liquid, thus causing earthquakes which also occur when the interior dries out and cracks
GEI 13

Fourth Century BC

ARISTOTLE (384-322 BC)

De Coloribus

Said the streak of a mineral revealed the true color of its constituent particles
BROME 116

De Generatione et Corruptione

Said the baser metals were changed into nobler metals over a long period of time
AD 297

Mechanica

Said elongate materials become broken and rounded when transported by running water
KRY 1722

Meteorologica

Said continuous, gradual erosion and silting caused major changes in physical geography
RUD 37-8

Said earth was round and relatively small
GEI 12

Said earthquakes were result of vapors created by moisture (rain) and heat (sun and interior) which flowed into, through and out of the earth and therefore earthquakes are commonest in countries where earth is porous and near the ocean, especially if shoreline is broken and cavernous
AD 407-8, 427-8; BROME 97

Said fossils were animals which had formed out of the moist earth and slime
HMPD 325

Said land and sea alternated locations
BROME 93; HMPD 325

Said meteoric stones had formed on earth, been carried aloft by wind, and subsequently had fallen back down
AD 19

Said sun´s heat caused water to evaporate and rise by day, and condense at heights or in cold mountain air, to fall as dew at night, or to form clouds and fall as rain when warm air is encountered and to enter the earth, only to consolidate and emerge as springs and rivers from the mountains which act as sponges
AD 27-30; BISWAS 66; GEI 29; OSR 4

Said the earth changed ceaselessly, albeit slowly, and nothing lasted forever
HMPD 325

Said the quantity of winds built up inside the earth determines the severity of earthquakes and, if they catch fire, volcanoes
AD 408

Said there had been a series of reversals between tropical and polar temperature regions and between dry land and seas
BEL 20; BROME 93; CHOR 6; GEI 34-6

Said stones were grown on or in the earth by the sun´s transmitted heat which also forms metals by mixing earth and water and, similarly, starlight grows gemstones
AD 17, 81, 278

Physics

Said matter, which consists of four properties or qualities, combined with form creates fire, water, earth and air which combine variously to form all else and exist in concentric zones (earth, then water, then air, then fire - the stars and planets encircling the earth)
AD 428; BJG 16; GEI 12; MHG 196

Respiration

Said fossils were remains of fish which lived in the earth
AD 12

EPICURUS (342-270 BC)

On Nature (37 books)

Said spring water was created deep inside the earth
OSR 4

c. 360 BC

PLATO (426/8-347/8 BC)

Critias

Said rain falling to earth was absorbed and directed into rivers and springs
BISWAS 66

Said soil erosion destroyed the hills, but warm, even temperatures of recent past had slowed the wearing away
KRY 1722-3

315? BC

THEOPHRASTUS (372?-287? BC)

De Lapidus (On Stones)

Classified materials as 1) earths, 2) metals, and 3) stones, and the latter according to appearance, physical characteristics, sources, and uses
AD 19, 21; HTOD 22

Classified minerals as 1) combustible, 2) incombustible, and 3) soluble in water and vitrifiable in fire
SMAJ 5, 13

Said coral grew in the oceans but in substance was like a rock
AD 130

315? BC

Said stones grew from seeds, light colored stones being male and darker stones being female
AD 95-6

On Fishes

Said fossil bones were produced by a plastic virtue in the earth
HMPD 325

Said marine fossils were the result of fish-eggs being caught up in the earth or were the remains of fish that wandered there and became petrified
GEI 16; HMPD 325

Said that there was a plastic force in the earth which formed imitation bones and plants
GEI 16

ZENO (c. 340-c. 265 BC)

Said fire is the primordial element from which all things have grown
HTOD 22

Third Century BC

PLATO (426/8-347/8 BC)

Cosmology

Said the earth was encircled by sun and planets and an outer ring of stars
ZIT 6

Dialogues

Said there were isosceles and equilateral triangles of matter which combined to form tetrahedrons (fire), cubes (earth), octahedrons (air), and icosahedrons (water), and thereby all materials of nature
BJG 14; MHG 196

Phaedo (and in Meteorologia)

Said a surging lake within the earth fed rivers, springs, and oceans
AD 427; OSR 4

Second Century BC

AGATHARCHIDES (or AGATHARCUS) (170-100 BC)

(Fragments only of his works remain)

Refers to gold as spreading itself through the earth´s crust like a tree
AD 289

First Century BC

LUCRETIUS (TITUS LUCRETIUS CARUS) (99?-55 BC)

De Rerum Natura

Said atoms had hooks and eyes and connected together, or interlocked to make larger elements
BJG 13-4

Said atoms of the four elements (fire, water, earth, and air) combined to form all matter
AD 33-4

Said earthquakes were caused by violent winds inside the hollow earth and also by the collapse of subterranean cavern walls
AD 33-4, 402-3

Said rising moisture formed clouds and then springs were filled by rain, but rivers were fed by ocean waters seeping through the earth from subterranean rivers and lakes
BAKER 398; BISWAS 96-7, 99; BJG 13-18; OSR 11

Said volcanoes resulted when winds trapped in the earth heated the rocks to magma
GEI 17

STRABO (63/65 BC-21/24 AD)

Geographica (Geography)

Said a volcano had uplifted a mountain
AD 331

Said deltas occur when river deposits are kept from the ocean by tides and are largest where rivers drain large areas of soft rock
GEI 30

Said most Mediterranean islands had been part of Italy until an earthquake split them off, although some islands were volcanic in origin
ZIT 8-9

Said rains caused floods but the source of all rivers was deep in the earth
FENT 26-7

Said sea level rises and falls in response to similar rise and fall of land masses
ZIT 8-9

Said the sea floor rises and falls affecting sea level and inundating the land, then retreating
BEL 20

Said uplift was caused by the earth´s inner fire
FENT 26

Said volcanoes were safety valves for fiery winds contained inside the earth which would otherwise cause earthquakes
AD 21, 26; FENT 26

VIRGIL (70-19 BC)

Aeneid

Said the interior of the earth contained a great lake from which issued the rivers
OSR 4

c. 15 BC

VITRUVIUS (MARCUS V. POLLIO) (First Century BC)

De Architectura

Said groundwater was a mixing of rain and snow seeping through the earth
CHOW 1-7

Said springs were created by rainwater and would be found wherever inclination of strata permitted accumulation of water
BAKER 398

Said the amount of discharge from a stream was equal to the cross-sectional area of the stream
BISWAS 89

First Century

HERO OF ALEXANDRIA (Fl. 65-150)

Dioptra

Said the discharge of water in a stream was equal to the cross-sectional area of the stream times the velocity of the water
BISWAS 86

63?

SENECA, LUCIUS ANNAEUS (3 BC-65)

Quaestiones Naturales

Said the four elements 1) air, 2) fire, 3) earth, and 4) water change one into the other so water running down mountains was air and that creating springs was earth
OSR 5

Said earthquake movements were 1) up and down, 2) oscillatory, or 3) vibratory
GEI 22-7

Said springs and rivers were fed by moist vapors rising from the center of the earth
AD 431; BAKER 399

Said the size of the subterranean chamber in which winds were trapped determined the size of the area to be affected by expansion and the resulting earthquake
GEI 22-7; ZIT 9

Said there are different waters in the earth some of which form metals and some of which harden into stone
AD 278

Said there were laws which governed all natural happenings
GEI 22

Said water in the earth´s center contains all four elements and rises, depositing metals in veins and becoming earth and stone and finally escaping into rivers
GEI 31

Said winds blowing through the earth heated combustible materials by friction, creating fire and volcanoes
GEI 22-7; ZIT 9

77

PLINY THE ELDER (GAIUS PLINIUS SECUNDUS) (23-79)

Historia Naturalis

Said water seeks the center of the earth and, when forced upward by some spirit, then flows through the earth´s veins to feed springs and rivers
BISWAS 99; OSR 4

Said coral was a shrub with berries, which hardens and turns red when taken out of the water
AD 130-1

Said crystals had definite shapes
WOOD 5

Said earthquakes occurred during calm periods when winds which had sunk into the earth broke forth again; suggested sinking ventilation shafts to effect peaceful escape
AD 413; GEI 26-7

Said geodes were stones in the act of giving birth
AD 36-45

Said some fossils fell during thunderstorms while others fell from the sky when the moon was eclipsed
HMPD 326

98?

FRONTINUS, SEXTUS JULIUS (35?-104)

De Aquis Urbis Romae Libri II.

Said stream discharge was determined by the cross-sectional area of the stream
BISWAS 89-90

First/Second Century

TERTULLIAN (155/160-220?)

Liber de Resurrectione Carnis (Concerning the Resurrection of the Flesh). Also see Liber de Pallio

Said shells found on mountains and hills had been deposited there during the flood
CAM 74; HMPD 326

c. 181

THEOPHILUS OF ANTIOCH (115-181?)

Apology of Christianity

Said the world was created in 5529 BC
MHG 196

Fourth Century

AMBROSE (339-397)

Said ocean waves tore down the land along its coasts and removed the debris from its bed to make room for increased waters
AD 332

LO HAN (Fl. 375)

Hsiang Chung Chi

Said fossils were once living organisms
HES 35

SEVERIAN OF GABALA (Fl. Late Fourth Century)

Said valleys were made for the Noachian floodwaters so there could be dry land again
AD 332

Fifth Century

ĀRYABHATA (476-550)

Āryabhatiyam

Said the earth moved through space with the other planets
MUR 52

c. 440

EUSTATIUS (Fifth Century)

Commentarius in Hexameron, in Patrologiae Cursus Completus, Series Graeca, by Jacques P. Migne

Said fossil fish had been deposited on mountains by the flood
CAM 74-93

Sixth Century

PROCOPIUS OF GAZA (465?-528)

Commentarius in Hexameron

Said fossil fish on mountains had been deposited there by the flood
CAM 74

VARĀHAMIHIRA (499/505-587)

Brhatsaṁhita (The Great Compendium). Madura; South Indian Press (1884)

Said geographical location determined in what part of the day an earthquake could occur (northwest, early morning, etc.)
ESSAI 34

Said the moon's location caused, and determined the severity of, earthquakes
KAR 97-8

Kōchit Bhuvah

Said earth was round and circled by planets
MUR 52

Said gems were rocks transformed by natural processes and time
MUR 52

Seventh Century

ISIDORE, SAINT OF SEVILLE (560-636)

Etymologiae Sive Origines

Said compacted air becomes clouds, and thickened produces rain, while frozen clouds produce snow
BISWAS 122

Said springs and rivers flowed from, and back to, the abyss inside the earth
AD 336

Eighth Century

ALCUIN (735-804)

Interrogationes et Responsiones in Genesin

Said primordial world was topographically smoother and the mountains were low
CAM 95

Ninth Century

RABANUS MAURUS (780?-856)

Commentaria in Genesin. Cologne; N.P. (1532)

Said the early earth had fewer topographic extremes
CAM 95

Ninth/Tenth Century

RHAZES (AL RAZI) (860?-925?)

Kitab Al-Hawi

Said there were six classes of minerals
MHG 196

c. 957

AL-MASŪDĪ, ABŪ-L-HASON ALĪ-IBN AL-HUSAIN IBN ALĪ AL MASŪDĪ (?-957)

Book of Indication and Revision

Said there was evolution from mineral to plant, from plant to animal and from animal to man
STN 639

1021-1023

AVICENNA (980-1037)

De Congelatione et Conglutione Lapidum, in Kitab al-Shifa (Book of the Remedy)

Said that metals are not transmutable but formed from different combinations of earths
AD 19; STN 709-13

Said that meteoric stones fall to earth from the heavens
AD 19

Said earthquake-induced uplift created some highlands, but mountains were cut from softer rock by winds and water
FENT 32

Said fossils were an attempt by nature to produce organic life out of the inorganic
AD 254

Said stone is formed through the hardening of clay, after precipitating from water through action of the mineral virtue, vis plastica
AD 83, 254, 333; ZIT 113

Classified minerals as 1) stones, 2) fusible substances, 3) sulphurs, and 4) salts
AD 19; HTOD 23

Said the sea once covered the earth completely
GEI 43

Said that the heights are uplifted by wind inside the earth and by erosion of surrounding softer land
AD 333

Said water is the main agent in changing topography
CHOR 6; GEI 43

Twelfth Century

AVERROES (MUHAMMAD IBN AHMAD) (1126-98)

Commentary on Aristotle

Said there was a smallest part into which a substance could be divided and still retain its identity
SCH 2

CHU HSI (1130-1200)

Chu Tsi Shu Chieh Yao

Said fossils were once living organisms
MHG 196

Said soft materials of the sea floor hardened and were uplifted to form mountains
BROME 119

Thirteenth Century

DUNS, JOHANNES SCOTUS (1266-1308)

De Rerum Principio

Said metals and stones were alive
BJG 20

1260

MAGNUS, ALBERTUS (1193-1280)

De Mineralibus et Rebus Metallicis Liber Quinque. Coloniae; Joannem Birckmannum and Theodorum Baumium

Said fossils were the result of a virtue formativa in the earth, although plant and animal remains might become petrified
AD 83

Said minerals and stone were generated by a formative mineral virtue when suitable constituents of the earth were combined with a favorable environment
AD 83

1282

D´AREZZO, RISTORO (Thirteenth Century)

La Composizione del Mondo di Ristoro d´Arezzo. Roma; Enrico Narducci (1859)

Said distant stars can draw earth above the water and raise mountains, with near stars creating valleys, although mountains may also be raised by earthquakes or were deposited by the Noachian flood
AD 335, 442

1282

Said the northern hemisphere had more land than the southern because there were more stars to draw the land above the waters
AD 335

Said molten interior of the earth could fracture and displace crust, thus raising or destroying mountains
GORT 504

1300

VĀGBHATA (Fl. 1280-1320)

Rasaratnasamuchchaya. Poona; Anandasrama (1890)

Classed minerals as 1) eight rasas, 2) eight uprasas, 3) gems, and 4) metals and their alloys
MUR 52

1320

ALIGHIERI, DANTE (1265-1321)

Quaestio de Aqua et Terra (A Question of the Water and of the Land). London; D. Nutt (1897)

Said dry parts of the earth's surface were drawn up out of the ocean waters by the attraction of the fixed stars or by the generating vapors they form in the earth
AD 342; OSR 7

Fourteenth Century

ALBERT OF SAXONY (ALBERTUS DE HELMSTEDE, ALBERTUS DE SAXONIA, etc.) (1316?-1390)

Said ocean would eventually cover the land as erosion lowered the surface but that local uplift compensated for the loss
STN 1429

1475

MEGENBERG, KONRAD VON (1309-74)

Buch der Natur (Book of Nature). Augsberg; Johann Bamler

Said earthquakes were the result of an accumulation of winds in the earth's interior forcing their way towards the surface
AD 404

c. 1500

VINCI, LEONARDO DA (1452-1519)

Notebooks of Leonardo da Vinci, edited by E. MacCurdy. London; Jonathon Cape/New York; Reynal and Hitchcock (various years)

Said of rivers: volume entering would equal volume leaving; the steeper the channel, the swifter the current and the faster the bed will erode; the longer water falls, the faster its current; surface water moves fastest because bed resistance exceeds that of air
BISWAS 144-5

Said fossils were remains of once living organisms
HMPD 327; RGIF 327

Said mountains were erosional remnants of aqueous action on the earth
AD 342

Said mountains were remnants of former earth's surface left after surrounding plains subsided
BEL 22

Said movement of the earth's crust was result of changes in load
LEEA 302

Said pervious beds of mountains, lying above impervious beds and dipping, carried rain and melted snow downward into rivers
AD 445-6

Said rivers eroded valleys depositing the detritus at the mouth, where plants and animals also were buried, petrified, and reappeared as fossils when the land rose or when seasonal flooding carried them inland
D&B 16; ZIT 23-24

Sixteenth Century

OLIVI OF CREMONA
(Sixteenth Century)

Said fossils were mere sports of nature
HMPD 328; ZIT 16

PARACELSUS (1493-1541)

Philosophy Addressed to the Athenians, in Opera Omnia Medico-Chemico-Chirurgica. Geneva; De Tournes (1658)

Said there were four elements: 1) air, 2) fire (producing day and night, cold and heat), 3) earth, and 4) water (producing all minerals and half of all nutrients)
MSSC 133

1502

LEONARDUS, CAMILLUS (Fl. 1502)

Speculum Lapidum Clarissimi Artium et Medicine. Venice; Melchior Sessam et Petrum de Rauanis Sociis (1516)

Said coral was a stone which grew, in sea-water, like a leaf-less shrub
AD 131

Said stones are produced by the focus of 1) God´s power, 2) stellar and planetary influences, and 3) elemental qualities (heat, cold, dryness, dampness, etc.)
AD 84, 155

Said the identification of stones depended on their hardness or softness, porosity, gravity, and density, among other things
AD 155-159

1504

REISCH, GREGOR (?-1525)

Margarita Philosophica (Second Edition). Argentinae; J. Schott

Said vapors rose through cracks in the earth, condensed, and emerged as springs
AD 433

1516

D´ANGHIERA, PIETRO MARTIRE (1455-1526)

De Orbe Novo (The Eight Decades of Peter Martyr). Alcala; A. Guillelmi

Said veins of gold were branches of a golden tree growing up from inside of the earth
AD 287

1517

FRACASTORO, GIROLAMO (1483-1553)

In Opera Omnia. Venice; Juntas (1555)

Said fossils were creatures which had lived where they were deposited when land and sea changed places
HMPD 327; RUD 41; ZIT 14-15

1518

KALBE, ULRICH RULEIN VON (?-1523)

Eyn wolgeordnet unt nutzlich Buchlin, wie man Bergwerck suchen und finden sol, von allerley Metall, mit seinen Figuren nach Gelegenheyt dess Gebirgs artlich angezygt mit anhangenden Berctnamen den anfahrenden Bergleuten vast dinstlich. Worms; N.P.

Said celestial rays mature base metals into nobler ones until gold is formed, after which, reduction to earth occurs
AD 298-301

1519

VINCI, LEONARDO DA (1452-1519)

Treatise on Water (never completed), in Notebooks of Leonardo da Vinci, edited by Richter, Jean Paul. New York; Dover (1970)

Said seawater rose to mountain tops, formed rivers, and returned to the ocean
BISWAS 140-141

Said the heat of the earth vaporized seawater which rose, cooled, and fell in the form of rain or snow
BISWAS 140-141

1522

ALEXANDER AB ALEXANDRO (1461-1523)

Geniales Dies. Romae; I. Mazochii

Said fossil shells settled on the sea floor which rose and concreted preserving the shells and forming mountains by succes sive depositions in the sea
RAM-2 6-7

1526/27?

PARACELSUS (1493-1541)

Book about Minerals

Said rivers were streams from the water-filled interior of the earth
MSSC 135

Said seeds of minerals and metals mature within the watery earth growing like plant
MSSC 134-5

Liber de Renovatione et Restauratione

Said the planets affected the early development of metals but the star which produces gold and silver could mend an imperfect metal
AD 284

1530

AGRICOLA, GEORGIUS (GEORGE BAUER) (1494-1555)

De Re Metallica. Misena; Frobenium (1551)

Said that minerals occur when a lapidifying juice deposits its mineral content, because of drying heat or freezing cold
CORB 8

Said varying proportions of earth, air, fire and water accounted for the differences in external characteristics of minerals
SMAJ 13

1534

SAVONAROLA, GIROLAMO (HIERONYMUS) (1452-98)

Compendium Totius Philosophiae tam Naturalis and Moralis. Venetiis; A. Pincii

Said stones grew everywhere: in caverns, quarries, fields, in oceans and lakes, in the air, and in the human body
AD 103

1538

FRACASTORO, GIROLAMO (1483-1553)

Homocentrica Sive de Stellis. Venice; N.P.

Said fossils were remains of once living animals
AD 261

De Natura Fossilium. Basel; N.P.

Classified mineral substances as 1) homogeneous earths, salts, gemstones, metals and other minerals; and 2) heterogeneous, i.e., rocks
AD 9; HTOD 24

Said minerals and fossils were organic as well as inorganic in origin
ZIT 17

Classified earth materials primarily by their physical properties as earths, rocks (compounds), metals, hardened fluids and stones (the latter included true fossils and gems)
RUD 23; SMAJ 13

Classified minerals according to origin, composition and appearance
FRMS 190

1538

<u>De Ortu et Causis Subterraneorum</u>. Basel; H. Frobenium et N. Episcopium

Said stones and minerals form by precipitation caused by heat or cold, from solutions of rain and sea water and inner-earth vapors moving through cracks in the earth
AD 93, 188, 309; FORB 28

Said mountains resulted from erosion of plains by water, assisted by earthquakes, subterranean winds, volcanoes, and winds piling up debris
AD 342-3

Said the earth's interior was heated by coal beds, bitumen, and sulphur which produced vapors which were themselves warmed by friction of gas against gas, and against walls of narrow spaces through which they moved rapidly, sometimes causing earthquakes when they could not escape otherwise
AD 281, 408-9

Said waters of the earth originate from rain, condensation of vapors and rivers within the earth, and from the ocean
AD 445

<u>De Re Metallica</u>. Misena; Frobenium (1551)

Said fossils were fatty matter which had been fermented by the earth's heat
HMPD 327

1550

BIRINGUCCI, VANNUCCIO (1480-1538)

<u>Pirotechnia</u>. Venezia; per Giouan Padoano, a Infantia di Gurtio di Nauo

Noticed the constancy of angles of pyrite faces
GORT 507

CARDANO, GIROLAMO (CARDANUS HIERONYMUS) (1501-76)

<u>De Subtilitate</u>. Nuremberg; I. Petreium

Said earthquakes were caused by hot vapors combined with nitre, bitumen and sulphur in earth's interior
AD 409

Said minerals were alive and grew like higher organisms, and metals grew from seeds, like plants
AD 94, 290

1552

De Subtilitate, Libri XXI. Basileae; L. Lucium (1554)

Said fossils proved the sea once covered the land
GEI 51-2; RUD 41

MATTIOLI, PIETRO ANDREA (1500-77)

Il Dioscoride dell'Eccellente Dottor Medico M. P. Matthioli. Vinegia; Vincenzo Valgrisi

Said fossils were formed by a 'Materia Pinguis' unlike bones and shells which were buried and then petrified
GEI 52; HMPD 328; ZIT 16

1554

VELCURIONIS, JOHN (Fl. 1553)

Ioannis Velcurionis Commentariorum Libri IIII, in Universam Aristotelis Physicem Nunc Recens Summa Fide Exactaque Diligentia Castigati et Excursi. Lugduni, I. F. de Gabiano

Said mountains originated from earthquakes, floods, wind, man, giants or God and were destroyed by these same natural means
TER 221

1557

CARDANO, GIROLAMO (CARDANUS HIERONYMUS) (1501-76)

De Rerum Varietate, Libri XVII. Basil; Henrichum Petri

Said vapor condensed into water and filled rivers and streams
BISWAS 148

Said the earth was full of water so surface waters could not disappear underground
BISWAS 148

FALLOPPIUS, GABRIEL (1523-62)

In Opera Genuina Omnia. Venice; N.P. (1584-1606)

Said fossil remains of elephants and sharks teeth were earthy concretions
GEI 52

Said fossils were formed in situ by fermentation
HMPD 328

1558

AGRICOLA, GEORGIUS (GEORGE BAUER) (1494-1555)

De Natura Fossilium, in Opera. Basel; H. Frobenio et N. Episcopioi (1563)

Said fossil shells were composed of earth and water and grew in the earth just as minerals and plants do
CAM 93

1558

Classified bitumen as 1) petroleum, 2) coal, 3) jet and obsidian, or 4) camphor and amber
FORB 28

1562

MATHESIUS, JOHANN (1504-65)

Sarepta. Nurnberg; J. von Berg

Said the heavy metals formed as a soft plastic and then hardened
AD 283

1564

FALLOPPIUS, GABRIEL (1523-62)

De Medicatis Aquis atque de Fossilibus. Venice; Jordanis Ziletti

Said dendrites and stone figures were result of fermentation in the rocks
AD 255

Said fluid circulating through the earth's crust 1) mixes with water to form porous rock, or 2) dissolves to form compact, solid rock
AD 90-1

Said mountains formed from exhalations which ascended through the earth, mixing with water to become stone, which lasted forever
AD 344

Said ocean water descending into the earth was evaporated by sulphur and bitumen fires, and rose to the surface where it condensed as pure spring water, leaving a deposit of salt below
OSR 12

MASSEI, GIOVANNI CAMILLO

Scala Naturale, Overo Fantasia Dolcissima. Venice; N.P. (1650)

Said the world consisted of four elements (earth, water, air and fire) and ten heavens (moon, sun, Mercury, Venus, Mars, Jupiter, Saturn, the stars, and the movements in the heavens)
OSR 6-7

1565

GESNER, CONRAD (1516-65)

De Rerum Fossilium, Lapidum et Gemmarum Maxime, Figuris et Similitudinibus Liber (A Book on Fossil Objects, Chiefly Stones and Gems, Their Shapes and Appearance), in Omni Rerum Naturae ac Philologiae Studiosis, Utilis et Jucundus Futuris. Zurich; N.P.

Classified minerals into fifteen categories based on their resemblance to animals, plants, shapes, etc.
FRMS 189

Said basalt crystallized from water
AD 127

Said minerals could be identified solely by visible characters
AD 175-83

Said some fossils were remains of plants or animals and some were formed by an inorganic process
AD 114; GEI 45

Said fossil figures formed simultaneously with the rocks
AD 257

1566

CESALPINO, ANDREA

De Re Metallicis. Rome; A. Zannetti (1596)

Said fossils were remains of once living creatures left behind by the sea and affected by the petrifying influences of the rocks
AD 261; GEI 53

1569

BESSON, JACQUES (Sixteenth Century)

L'Art et Science de Trouver les Eaux et Fontaines Cachées sous Terre. Orléans; P. Trepperel

Said water was evaporated by the sun, and then came down as rain which entered springs and rivers and returned to the sea
BISWAS 157

1571

GALESIUS, AUGUSTINUS

Terrae Motu Liber. Bologna; A. Benaccium

Said earthquake movements are 1) vibration, 2) tremor, 3) upward, 4) downward, and 5) pulsating up and down
AD 405-8

1571

Said earthquakes are the result of the action of fire, air, and water in the earth, which generates exhalations which then force their way to the surface
AD 405-8

1574

MERCATI, MICHELE (1541-1593)

Metallotheca Vaticana. Rome; Ex Officiona J. M. Salvioni (1717)

Said figured stones, fossil fishes, and dendritic growths were caused by radiation from heavenly bodies
AD 253; GEI 52; HMPD 328

1575

AUBERT, JACQUES (?-1586)

De Metallorum Ortu et Causis Contra Chemistas Brevis et Dilucida Explicatio. Lyons; Iohannem Berjon

Said that metals were not alive and did not grow from seeds
AD 291

FRASCATUS, GABRIEL (1520?-1581)

De Aquis Returbii Ticinensibus Commentarii Mineras. Ticin; Hieronymum Bartholum

Said heavenly rays penetrate and heat earth´s interior thus maturing metals whose quantity and purity depend on the amount of heat received
AD 279-80, 297

BACCIUS, ANDREAS (1550-98)

De Gemmis et Lapidibus Pretiosis. Frankfort; N. Steinii (1603)

Said gems and gold are created when the heavenly powers of the stars focus on the earth
AD 84

1578

BOURNE, WILLIAM (?-1583)

Booke Called the Treasure for Traveilers. London; T. Woodcocke

Said flood waters eroded the banks of rivers thus creating marshes inland and sandbanks downstream
GLD 257

Said sea cliffs are eroded by the waves which then reduce and round the debris
GLD 259

1579

MATTIOLI, PIETRO ANDREA (1500-77)

Commentaires sur les Six Livres de Ped. Dioscorides (Commentaries on the Six Books of Dioscorides). Lyon; Guillaume Rouille

Said there were fluid bodies circulating through the earth's crust which crystallized into various gems

AD 92

1580

PALISSY, BERNARD (1499-1589)

Discours Admirables de la Nature des Eaux et Fonteines, tant Naturelles, Qu'artificielles, des Métaux, des Sels et Salines, des Pierres, des Terres, du Feu et des Emaux. Paris; Chez Martin

Said fossil shells and fishes once were living organisms
HMPD 328

Said a salt in water attracted specific materials, reducing them to metals
MSSC 136-7

Said an oil in water congealed, and thus metals formed and sulphur was released
MSSC 136-7

Said evaporation produced rainwater, which then flowed downward to emerge as springs and rivers, but that artesian springs were fed by waters originating above them
AD 446-8; BISWAS 153; BROME 125; OSR 12

Said extinct marine species had lived inland in salt lakes
RUD 42

Said fossils were once living, some now extinct, creatures hardened by mineral deposition from congelative waters
AD 261; BROME 125; EDW 21; PPHG 5

Said the Parisian region had once been covered by water
BROME 125

Said pyrite crystals grew from liquid seeds in clay beds
AD 290

Said rocks began as clays
AD 446-8

1580

Said the flood could not be dated from fossils found in rocks
AD 446-8

1583

ARETINO, GIRL. (GIROLAMO?) BORRO (1512-92)

Des Flusso e Reflusso del Mare e dell´Inondatione del Nilo. Firenzi; Georgio Marescotti

Said the narrowness of the fissures through which water traveled through the earth filtered out salt particles
OSR 11

1584

BRUNO, GIORDANO (1548-1600)

Cena de le Ceneri (The Ash Wednesday Supper). London; J. Charlewood

Said there were frequent alterations in the distribution of land and sea
ZIT 24

De Immenso et Innumerabilis; Seu de Universo et Mundis (On Immensity). London; J. Charlewood

Said thermal and volcanic phenomena might be caused by the reaction of surface waters entering the earth´s interior
ZIT 24

Said the earth is a spherical body whose oceans are deeper than its mountains are high
ZIT 24

Said thermal and volcanic phenomena might be caused by the interaction between surface waters and the interior of the earth
ZIT 24

Spaccio dela Bestia Trionfante (The Explusion of the Triumphant Beast). London; J. Charlewood (1713)

Said there never was a universal deluge, only local flooding
ZIT 24

1586

CAMDEN, WILLIAM (1551-1623)

Britannia. London; Radulphum Newberry

Said ammonites were a product of nature
COX 14

1590

ROY, LOUIS LE (c. 1510-77)

Of the Interchangeable Course of Things in the Whole World. London; C. Yetsweirt (1594)

Said rivers and springs dry up, ocean and land change places, and mountains become plains which uplift and become mountains
TER 221-2

1592

AQUINAS, THOMAS (1224/25-1274)

De Esse et Essentia Mineralium. Cologne, N.P.

Said metals were formed by rays from the planets, moon and sun: Saturn forms lead, Venus forms copper, the moon forms silver, and the sun forms gold, etc.
AD 282

1599

IMPERATO, FERRANTE (1550-1631)

Historia Naturale. Napoli; C. Vitale

Said fossils had been animals and were generated by water and humidity
EDW 22

c. 15??/16??

FORSIUS, SIGFRID ARON (1550?-1624)

Minerographia. Stockholm; Ignatius Meurer (1643)

Said that amber is petrified spruce or pine resin
PPHG 9

1600

GILBERT, WILLIAM (1540-1603)

De Magnete, Magneticisque Corporibus et de Magno Magnete Tellure. London; Petrus Short

Said crystals were water solidified by extreme cold or some other nature of the soil
BJG 23

Said earth´s interior was largely iron, i.e., a lodestone
BRUSH 706; P&S 671

1601

LIBAVIUS, ANDREAS (1550-1616)

De Petroleis, Ambra, Halosantho, Succino, Gagate, Asphalto, Pissasphalto, Mumia, Lithantrace. Frankfort; P. Koff

Said amber was fossil resin and lithantrace was fossil bitumen
FORB 28-9

1603

OWEN, GEORGE (1552-1613)

First Booke of the Description of Penbrookshire in Generall, in Cymmrodorian Record Series, Number 1, London (1892)

Said the flood divided the land into parts and created the present topography
GLD 255

1604

SENDIVOGIUS, MICHAEL (1566-1646)

De Lapide Philosophorum. Frankfurti; Joannis Bringeri (1611)

Said moist subterranean vapors carried metal and mineral seed up into the earth where resultant species depended on the environment where it grew
MSSC 138-9

Tractatus Duodecim de Lapide Philosophorum. Prague; N.P.

Said water flowing through the earth carried seeds of substances and also kept the planet from burning up
MSSC 140

1605

ALDROVANDUS, ULYSSIS (1522-1605)

Geologia Ovvero de Fossilibus (posthumous). In Musaeum Metallicum. Bologna; I.B. Ferronii (1648)

Said corals on the sea bottom are woody, but the lapidifying juice present in sea water petrifies them
AD 131

In Musaeum Metallicum in Libros III Distributum

Classified minerals according to origin, nature, mode of occurrence, mythology, uses, etc.
FRMS 189

LYDIAT, THOMAS (1572-1646)

Praelectio Astronomica; de Natura Coeli et Conditionibus Elementorum; Tum Autem de Causis Praecipuorum Motuum Coeli et Stellarum. London; Ioannes Bill

Said floods were caused by excess rain (Noachian had been waters sent by God) and the exchange of oceans and land resulted from natural motion of the seas and subterranean fires
TER 223

1605

STEVIN, SIMON (1548-1620)

Hypomnemata Mathematica. Lugduni Batavorum; Ioannis Patii (1605-8)

Said earth´s surface elements were constantly transforming and changing position
FRANG 246

VERSTEGEN, RICHARD (1548-1640)

Restitution of Decayed Intelligence. Antwerp; R. Bruney

Said the continental coastlines were originally smooth but had been eroded by earthquakes and the flood, and especially by the waves
GLD 260

1609

BOODT, ANSELMUS BOECE DE (1550-1634)

Gemmarum et Lapidum Historia. Hanoviae; Typis Wechelianis apud C. Marnium and Heredes J. Aubrii

Classified minerals according to several considerations, such as beauty and place of origin
FRMS 190

Said corals died and then petrified
AD 132

Said gems have no magical powers but do focus heavenly powers
AD 161

1614

PAREUS, DAVID (1548-1622)

In Genesin Mosis Commentarius. Geneva; Petrum et Jacobum Chouet

Said the deluge produced mountains on an otherwise round earth
CAM 95

1616

COLONNA, FABIO (1576-1650)

Osservazioni Sugli Animali Aquatici e Terrestri (Observations on Some Aquatic and Terrestrial Animals), in Purpura. Romae; J. Mascardum

Said some fossils were once living creatures and described them biologically, distinguishing freshwater from marine
GORT 507; RUD 42; ZIT 19

1617

LOHNEYSS, GEORGE ENGELHARD VON (1552-1622)

Bericht von Bergwerk. Zellerfeldt; Gedruckt

Said God had planted seeds so metals, etc., would grow in the earth ... perpetually
AD 307

1618

SENNERTUS, DANIEL (SENARTUS) (1572-1637)

Epitome Naturalis Scientiae. Wittberg; C. Heiden

Said the weight of the ocean's waters, especially at high tide and in storms, forced it to flow into shoreline fissures and up into mountains where it emerged as rivers
AD 442; OSR 10

1619

KEPLER, JOHANN (1571-1630)

Harmonices Mundi. Lincii Austriae; Sumptibus G. Tampachii, Excudebat I. Plancus

Said earth constantly decomposed fluids and matter to form other materials
BISWAS 167

Said earth produced ground water and springs from absorbed sea water
BAKER 399

1620

BACON, FRANCIS (1521-1626)

Novum Organum, in Section XXVII of Aphorisms, Book Two. London; Johannem Billium

Observed the similarity between the African and Peruvian coasts
CAR-2 283; CDBN 349

1625

CARPENTER, NATHANIEL (1565-1628)

Geography Delineated Fourth in Two Bookes. Oxford; H. Cripps

Said mountains are continually wearing down and new ones are building up
TER 223-4

Said the height of a mountain was determined by its resistance to weathering by rain and snow
GLD 257

1625

Said early earth was a water-covered spheroid and most topography was determined during the creation when land and water separated
GLD 254; TER 216, 223-4

Said the earth would again be water covered, as it originally was
GLD 257

1628

CASTELLI, BENEDETTO (1577-1644)

Della Misura dell'Acque Correnti. Rome; Nella Stamperia Camerale

Said water in like parts of a river flowed in equal amounts at the same time but in unequal parts, the amounts will be proportionate
BISWAS 172

1631

JORDEN, EDWARD (1569-1632)

Discourse of Naturall Bathes and Minerall Waters. London; T. Harper

Said that minerals grew from seed forms which determined their ultimate shape and the heat released by this fermentation created hot springs
MSSC 146-7; STP 111

Said the interior of the earth was a solid core and volcanoes and heat-generated metals were produced when near-surface combustibles were ignited by lightning
STP 119-20

1633

DESCARTES, RENE (1596-1650)

Le Monde, in Le Monde de Mr. Descartes (posthumous). Paris; Jacques le Gras (1664)

Said the gravitational attraction of a small star collected matter which condensed, cooled and formed the planets
SCH 7

1635

CLAVE, ETIENNE DE (Seventeenth Century)

Paradoxes. Paris; Chez la Veufue Pierre Chevalier

Said rain falls, seeps through earth, takes up minerals, is volatized by the earth's central fire and rises, depositing mineral seeds which will grow
AD 87

SWAN, JOHN (Fl. 1635)

Speculum Mundi. Cambridge; Roger Daniel (1643)

Said earthquakes raised, destroyed, and moved mountains, submerged islands, and created new rivers, volcanoes, and straits
GLD 256

1636

MERCATOR, GERARDUS (1512-99)

Historia Mundi (Mercator's Atlas). Amsterdam; H. Hondius

Said a great wind formed the continents and ocean basins on the third day of creation
GLD 254

1637

DESCARTES, RENE (1596-1650)

Les Météores, in Discours de la Méthode. Leyde; J. Maire

Said sea water entered the earth, was evaporated by central heat, rose to the mountain peaks, condensed, and emerged as rivers and streams
BISWAS 169-70

STELLUTI, FRANCESCO (1577-1646)

Trattato del Legno Fossile Minerale. Roma; V. Mascardi

Said lignite was earth which slowly became woody from the effect of subterranean heat and sulfur water
EDW 5

1639

PLATTES, GABRIEL (Fl. 1638)

A Discovery of Infinite Treasure, Hidden Since the Worlds Beginning. London; I. Legatt

Said earth was held in the center of the universe by the magnetic attraction of the heavenly bodies
DEBUS 162-5

Said in addition to the flood, there was a submergence during which the present topography was water-cut
GLD 260

Discovery of Subterraneall Treasure. London; Jasper Emery

Said bituminous and sulfurous vapors created rocks and mountains, and metals filled in the remaining cracks
DEBUS 165

1639

Said great heat was required to form gold and silver so there was less and less as you approached the poles
AD 83

Said transmutations by minerals could happen naturally (iron plus copper into gold, for example)
DEBUS 163

1640

BARBA, ALVARO ALONSO (1569-?)

Arte de los Metales (Art of Metals). Madrid; Imprenta del Reyno

Said ore deposits grow continually
AD 306-7

1644

DESCARTES, RENE (1596-1650)

Principia Philosophiae. Amstelodami; Ludovicum Elzevirium

Said exhalations rise through earth expanding and burning to escape as earthquakes and/or volcanoes
GEI 80-1

Said planets formed from whorls of fiery star-matter which then cooled from the surface inward and coagulated into density layers according to the individual gravities of their components
FRANG 345; GEI 80-81; HAZ 188; HER 9

Said veins occurred when a layer of heavy minerals, heated in the earth's fiery interior, released exhalations which rose through fissures and cooled and solidified
HTOD 27; TODH 312

LIGHTFOOT, JOHN (1602-75)

Harmony of the Foure Evangelists, Part 1. London; R. Cotes for Andrew Crocke

Said creation occurred in 3928 BC
WRB 19

1646

BROWNE, THOMAS (1605-82)

Pseudodoxia Epidemica. London; T. H. for Edward Dod

Said shape of a crystal, solidified by cold, was result of an inherent geometric principle within the mineral fluid
BJG 24

1647

DESCARTES, RENE (1596-1650)

Principes de la Philosophie. Paris; Michel Bobin and Nicolas le Gras (1668)

Said light rock layer of earth buckled and rose up to form mountains pocketed with caves full of air and water
UHS 384

Said subtle matter of heaven formed core of earth and was encircled by layers of iron, heavy rock, light rock, water, and lastly, air
UHS 384

PAPIN, NICHOLAS (?-1653)

Origin of Fountains, in Raisonnemens Philosophiques Touchant la Salure, Flux et Reflux de la Mer. Blois; Francois de la Savgere

Said the oceans were higher than continent so water were forced inland through the earth to emerge as springs and rivers, there not being sufficient rain and snow t fill them all
AD 441; OSR 10, 13

1648

HELMONT, JEAN BAPTISTE VAN (1577-1644)

Ortus Medicinae. Amsterdam; Luduvicum Elzevirium

Said fermenting waters produce both the seeds of metals and minerals and the mineral juices in which they develop
MSSC 140-3

Said plants grew by transforming water into solid vegetation
FRANG 350

1649

GASSENDI, PETRI (PIERRE) (1592-1655)

Syntagma Philosophiae Epicuri. Hagae-Comitis; A. Vlacq (1659)

Said atoms form different molecules which then act as seeds for the different substances
BJG 30

1650

SENDIVOGIUS, MICHAEL (1566-1646)

New Light of Alchymy. London; R. Cotes

Said metal deposits are the result of action of celestial rays on distillations and sublimation inside the earth
STP 103

VARENIUS, BERNHARD (1622-50)

Geographia Generalis, in Qua Affectiones Generales Telluris Explicantur. Amstelodami; L. Elzevirium

Said the most numerous and violent earthquakes occur when ignited nitre and sulphur vapors cannot escape through fissures, as is the case with volcanoes
AD 410; ZIT 25

1650-54

USSHER, JAMES (1581-1656)

Annales Veteris et Novi Testamenti. London; J. Crook

Said the earth was created in October of 4004 BC
SIMP 260; P&S 65; WRB 20

1651

GLAUBER, JOHANN RUDOLF (1604-68)

Operis Mineralis. Amsterodami; Joannem Janssonium

Said heat and high pressure in earth's interior propelled vapors toward the surface where they became water and metallic ores
AD 286

Said all metal constantly transformed to ultimately become gold
AD 286

De Ortu et Origine Omnium Metallorum and Mineralium, in Operis Mineralis. Amsterodami; Joannem Janssonium

Said a mineral deposit grew upward from the center of the earth, with branches, like a tree
AD 281

Said empty center of earth received rays from the stars which then rose surfaceward to become metal ores
AD 281

1656

BARATTIERI, GIOVANNI BATTISTA (1601-77)

Architettura d´Acque. Piacenza; G. Bazachi

Said water was distilled in the mountains and, joined by melting snow and rain, formed springs and rivers
AD 439

BLOUNT, THOMAS (1618-79)

Glossographia. London; Thomas Newcomb

Said bitumen was a clay or slime which grew
FORB 28

1657

DOBRZENSKY, JACOBUS JOHNNES WENCESLAUS (1659-84)

Nova et Amoenior de Admirando Fontium Genio. Ferrariae; Alphonsum et J.B. de Marestis

Said subterranean condensation and the tides create rivers
BISWAS 198

1660

DUHAMEL, JEAN BAPTISTE (1624-1706)

De Meteoris et Fossilibus. Paris; Petrum Lamy

Said water holding rocky materials in suspension produces stones
AD 91

1661

BECHER, JOHANN JOACHIM (1635-82)

Natur Kündigung der Metallen. Frankfurt; in Verlegung Johan Wilhelm Ammons und Wilhelm Serlins (1705)

Said lead-carrying water descended through the earth and encountered rising salt and sulphur vapors whereupon heavenly rays solidified the mixture in veins
AD 288-9

BOYLE, ROBERT (1627-91)

Sceptical Chymist. London; J. Cooke

Said earth has a plastik principle which may eventually change it to metal
AD 292-3

Said minerals developed from some sort of seed within the earth
MSSC 153

1661

JONSTON, JOHAN (1603-75)

Notitia Regni Mineralis. Leipzig; Jacobi Trescheri

Classified minerals according to appearance
FRMS 192

1662

LEFEVRE, NICHOLAS (1610?-69)

A Compendious Body of Chemistry. London; T. Davies and T. Sadler

Said star light reaches earth and is nurtured in fermenting waters where it matures into a mineral or metal
MSSC 148-50

1663

SCHOTTUS, GASPAR (1608-66)

Anatomia Physico-Hydrostatica Fontium ac Fluminum. Herbipoli; Jobus Hertz

Said some springs and rivers are fed from condensation of vapors rising through the earth, and some from rain and snow, but the majority are fed from water migrating through the earth
AD 443; BISWAS 180-1

1664

BECHER, JOHANN JOACHIM (1635-82)

Institutiones Chemicae Prodromae. Amsterdam; Elizeum Weyerstraten

Said sulphuric vapors mixed with mercury, which determines shape, to produce metals in the earth
MSSC 147

1664-65

KIRCHER, ATHANASIUS (1602-80)

Mundus Subterraneus (Subterranean World). Amsterdam; J. Janssonium and E. Weyerstraten

Said fossils and crystals were result of spiritus plasticus, a magnetic force, but larger bones were from the giants and rocks were the results of the lapidifying virtue of the earth
AD 255-6, 433; RISE 258, 260; ZIT 24

Said heat drove magma in the earth´s center up into volcanoes
BAKER 399

Said mountain ranges ran north-south, and a few east-west, due to the tides
AD 411; FRANG 351

Said openings on the sea floor admitted water to the earth's interior where it was heated by subterranean fires and was then drawn up into springs and rivers
AD 433; BAKER 399; BISWAS 178-9

Said rock-bound forms of leaves, fish, etc. were once living things
AD 255-6

Said sulphurous and nitrous vapors mixed within the earth exploded, causing earthquakes
AD 411

Pyrologus, in Book III of <u>Mundus Subterraneus</u>

Said there were innumerable subterranean centres of conflagration and they are connected with active volcanoes
ZIT 24

Said water cavities in the earth are fed by the sea and emerge at the earth's surface as rivers and springs
AD 433; ZIT 24

1665

HOOKE, ROBERT (1635-1703)

<u>Micrographia, or Some Physiological Descriptions of Minute Bodies Made by Magnifying Glasses, with Observations and Inquiries Thereupon</u>. London; J. Martyn and J. Allestry

Said fossils were shells which filled with mud, clay or petrifying water and then hardened
STOKES 16

Said crystals were composed of contiguous spheres equal in diameter but capable of forming geometric shapes when stacked in various numbers in a liquid matrix
BJG 39-40; SMAJ 11

Said fossil shells, hardened clay, or mud impressions of once living organisms, when found inland, were the result of a flood or earthquake
EPBG-I 130; RUD 54

Said petrifying water turned wood and shells to stone
EPBG-I 130; RUD 54

1666

BOYLE, ROBERT (1627-91)

Origine of Formes and Qualities (Essay about the Origin and Virtues of Gems). Oxford; H. Hall for R. Davis

Said that the earth molded mineral deposits that built up
MSSC 153

Said the universal matter set in motion at creation was divided into particles of varying sizes and shapes which aggregated into stone, metal, etc.
BJG 30

LANA, FRANCESCO TERZI (1631-87)

On the Formation of Crystals, in Royal Society of London, Philosophical Transactions, 7: 4068-69

Said quartz crystals were formed by the coagulation of dew and nitrous material
AD 132-3

1667

STENO, NICOLAUS (1638-86)

Canis Carchariae Dissectum Caput. Elementorum Myologiae Specimen, Seu Musculi Descriptio Geometrica. Florence; Ex Typographia sub Signo Stellae

Said the so-called tongue stones were actually the teeth of fossil sharks
RUD 50-2

Said strata had been deposited horizontally, in water, and later collapse of the upper strata due to erosion or inner heat of the earth accounted for tilt
BEL 23; SCHERZ 102

Said the earth was layered and full of fossil remains of dead animals
RUD 50-2; SCHERZ 28, 100

1668

HOOKE, ROBERT (1635-1703)

Discourse of Earthquakes, in R. Waller, Posthumous Works of Robert Hooke. London; S. Smith and B. Walford (1705)

Said earth's centre of gravity and axis of rotation might be affected by large earthquakes and altered conditions might then result in changes in species
GEI 70; RAN 323

Said disasters and mutations recorded in the earth could be dated by fossils
DEAN 448

1668

Said earlier climates had been warmer
GEI 70; RGIF 329

Said fossils could be used to develop a chronology
MHG 197

Said fossils were organic and could be useful in chronologic work
CBS 215

Said lighter parts of the earth rise as the heavier parts sink
RAN 322-3

Said that a succession of earthquakes could elevate land and uplift mountains
RAN 323

Said that a very strong earthquake might alter the earth´s center of gravity and ax of rotation thus affecting the time required to circle the sun, i.e., the length of the earth´s day
GEI 70

Said earth is constantly changing as the heights wear down and water circulating from the ocean, as rain, creates unconformities and carries detritus, in rivers, back to the ocean where deposition occurs again
DAVIES 495; RAN 322

Said fossiliferous strata took a long time to form
RAN 324

Said that fossils were organic, anatomical identifiable, and might be useful for making chronologic comparisons with rocks of similar age
D&B 19; RAM 1 18

Said that the effects of volcanoes and earthquakes balanced those of erosion and caused the land and sea to change places
ALB-2 47

Said there had been more subterranean fuel in the past so volcanoes and earthquakes were more frequent and severe
RAN 323

Said fossils indicated that the climate had been warmer in the past
JONES 245

Said fossils were deposited on mountains by volcanoes
JONES 245

Said that England had been warmer and that the ancient climates might have been different because the tilt in the earth's axis was different
WOOD 12

Said earthquakes and volcanoes were caused by accumulations of sulphurous vapours deep in the earth
RHRE 86

Said earthquakes buried fossils
PORT 83

Said earthquakes destroyed mountains and turned strata over, thus exposing the previously buried portions
BEL 25

Said some ancient species were extinct and some present species were new
RAM-2 19; RAN 322; RGIF 329

Said strata became consolidated by pressure of overlying strata, crystallization of components, and heat from the earth's interior
DAVIES 498; RAM-2 18

Said that land is uplifted, often bearing marine fossils, by earthquakes and other land then sinks accordingly
AD 421; RHREG 85-6

Said that some fossils might be extinct species which occupied small areas and were destroyed during earthquakes
RAM-1 19; ZIT 19

PLACET, FRANCOIS

La Corruption du Grand et Petit Monde. Paris; Alliot

Said that before the flood, there was one large landmass, but parts of it sank becoming ocean and causing other parts to rise higher
CAR-2 284; HAL 1

1668?

HOOKE, ROBERT (1635-1703)

Discourse of Earthquakes, in R. Waller, Posthumous Works of Robert Hooke. London; S. Smith and B. Walford (1705)

Said continents had moved, thus changing past climates
RAN 323

1669

BARTHOLINUS, ERASMUS (1625-98)

Experimenta Crystalli Islandici Disdiaclastici Quibus Mira et Insolita Refractio Detegitur. Hafniae; Sumptibus Danielis Paulii

Described double refraction of light in the islandic calcite
GRA 35; SMAJ 8

BECHER, JOHANN JOACHIM (1635-82)

Physica Subterranea. Leipzig; J. L. Gleditocchium (1703)

Said all materials formed from combinations of earth and water acted upon by heaven and fire
SPMS 271-2

Said original chaos yielded a ball of earth a layer of water and, surrounding it all, a layer of air
SPMS 271

Said veins were filled by vapor rising from the earth´s interior
TODH 314

LACHMUND, FRIEDRICH (1635-76)

Oryktographia Hildesheimensis, Sive Admirandorum Fossilium, Quae in Tractu Hildesheimensi Reperiuntur Descriptio, Iconibus Illustrata .... Hildesheim; Jacobi Mulleri

Said fossils were petrified natural objects or were formed by plastic virtues
CAM 93

STENO, NICOLAUS (1638-86)

De Solido intra Solidum Naturaliter Contento Dissertationis Prodromus (On the Solid Naturally Contained in a Solid). Florence; ex Typographia sub Signo Stellae

Said fossils were once living marine plants or animals whose bodies were replaced by minerals and that wherever marine fossils are found, there was once ocean and wherever land fossils are found, there was once a torrent of water
GEI 54-6

1669

Said all solids form from a liquid (air, water, or organic) and precipitation is an ongoing process
ALB-2 32; FENT 36

Said natural crystals could not be distinguished from artificially grown ones
RUD 60

Said of a body within another body, whichever hardens first will leave its impression on the other
ALB-2 32

Said strata are laid down horizontally on older beds
GEI 56-60; MARG 98

Said quartz crystals all have the same angle between their corresponding faces
ALB-2 36; P&S 112

Said similar compositions indicated a series of strata formed in an environment undisturbed by diverse fluids and sediments over time, and vice versa
GEI 54-6

Said solid bodies which are alike externally and internally probably originated similarly
ALB-2 32

Said strata are laid down horizontally, by settling through water onto older bottom layers encircling the earth until contained by a solid body
BEL 22-3; CORB 10-11

Said the history of the earth's development could be read in the rocks
GEI 56-60

Said the major agent of topography was running water
GEI 56-60

Said the presence of coals, ashes, pumice, and bitumin indicated the likelihood of subterranean fires in the past
GEI 54-6

Said the tilting of once horizontal strata, sometimes accompanied by uplft, produced hills and some mountains, and occurred when gases, or air, escaped the earth´s interior or downfall filled fissures
AD 358-60; GEI 56-60; SCHERZ 111; UHS 384-5

Said veins formed when minerals condensed out of vapors which rose from the earth´s fiery interior
TODH 312

Said volcanoes result from the combustion of buried carbonaceous material
GEI 56-60

Prodromus, in De Solido intra Solidum Naturaliter Contento Dissertationis Prodromus

Said crystals grow outward from around seeds and the shape of the crystal will be determined by the speed of growth in each direction
BJG 55; MSSC 153

Said some fossils were once marine animals whose bodies were replaced by minerals
RAM-1 12-4

Said strata are consolidated by the earth´s volcanic heat and are later tilted by upheavals and depression
RAM-1 12-14

Said the earth had gone through six periods of formation: 1) horizontal unfossiliferous strata deposited under the waters, 2) sea withdrew, land dried and cavities formed in the earth, 3) fire and water ate into the crust, breaking it into mountains and valleys, 4) waters returned and fossiliferous strata formed, 5) sea again withdrew and much erosion by rivers occurred, and 6) topography was created by fire, water, and collapse
BEL 23-4; ROD 239-41

1670

SCILLA, AGOSTINO (1639-1700)

La Vana Speculazione Disingannata del Senso. Napoli; Appresso a Colicchia

Said fossil aggregations had been deposited by a flood
HMPD 330; RAM-2 12-3

Said non-fish fossils had formed from juices or vapors which originated in the earth's interior
RAM-2 12-3

Said that fossil shells had been living marine creatures
HMPD 330; RAM-1 12-3

1671

LISTER, MARTIN (1638-1712)

Fossil Shells in Several Places of England, in Royal Society of London. Philosophical Transactions, 6: 2281-84

Said different stone formations produced different fossils and the rocks could be so identified
EPBG-1 131-2; GEI 336; ZIT 17

Said fossils were cockle-like stones of the same composition as the rocks surrounding them and never were part of any animal
AD 259; CONK 8; EPBG-1 132; WOOD 12; ZIT 17

WEBSTER, JOHN (1610-82)

Metallographa, or An History of Minerals. London; by A.C. for W. Kettilby

Said metals developed from seeds within cavities of the earth in the presence of salt water which had descended and been enriched by rising vapors
MSSC 150-1

Said metals, like plants, grew from seeds
AD 196, 291

1672

BOYLE, ROBERT (1627-91)

Essay about the Origin and Virtues of Gems. London; W. Godbid

Said crystals were composed of parallel platelets along which cleavage would occur
SMAJ 8

Said some liquid, possibly steam, consolidated soft earth into opaque gems and liquids into transparent gems
MHG 198

1672

JOSSELYN, JOHN (1608-75)

New Englands Rarities Discovered. London; J. Widdowes

Said sea waters flowed up into mountains, were desalted, and emerged as ponds and lakes
JJGO 173

Said valleys were cut by running water
JJGO 173

SHERLEY, THOMAS (1638-78)

Philosophical Essay Declaring the Probable Causes Whence Stones Are Produced in the Greater World. London; W. Cademan

Said vapors rise from the earth´s interior carrying seeds which water, and the fermentive odors present, cause to grow into stones and metals, which, in turn, procreate, as plants do, gold giving birth to gold, etc.
AD 87-9, 290

1673

RAY, JOHN (1627-1705)

Observations Topographical, Moral, and Physiological. London; J. Martyn

Said fossils were organic in origin
STOKES 19

Said that for mountains to have formed, the world had to be very old or to have been vastly rearranged through time
STOKES 19

1674

JOSSELYN, JOHN (1608-75)

An Acount of Two Voyages to New England. London; G. Widdowes

Said water flowed, and trees of metal grew upward, in the hollows of mountains
JJGO 179-80

Said earthquakes raised some mountains
JJGO 179-80

PERRAULT, PIERRE (1608-80)

De l´Origine des Fontaines. Paris; Chez Pierre le Petit

Said the rain filled rivers and springs
BISWAS 212-3

1676

BEAUMONT, JOHN (?-1731?)

Concerning Rock-Plants (Crinoids) Growing in the Lead Mines of Mendip Hills, in Royal Society of London. Philosophical Transactions, 11: 724-42

Said crinoids were stone plants with roots, stems, and branches produced by the power of nature to express the shapes of plants and animals
AD 257

Said that fossils grew like crystals
STOKES 18

1677

HALE, MATTHEW (1609-76)

Primitive Origination of Mankind. London; W. Shrowsbery

Said fossils found on mountains were either left by the flood or developed there when the flood spread seminal fluid found in the sea
CAM 93

LEMERY, NICOLAS (1645-1715)

Principes (Cours) de la Chymie. Paris; L'Auteur

Said basic constituents of nature are water, spirit (mercury), oil (sulphur), salt, and earth
MSSC 152-3

Said stones and crystals form from interaction of basic constituents plus subterranean heat or cold
MSSC 152-3

PLOT, ROBERT (1640-96)

Natural History of Oxfordshire. Oxford; The Theater

Said fossils were random shapes produced by a plastic virtue in the earth
AD 259; HMPD 328; RGIF 328; WOOD 13

1678

HERBINIUS, JOHANN (1633-79)

Dissertationes de Admirandis Mundi Cataractis Supra et Subterranis. Amsterodami; Janssonio Waesbergios

Said the waters in the interior of the earth were forced surfaceward to emerge as springs and rivers
AD 440-1; OSR 10-1

LISTER, MARTIN (1638-1712)

Historiae Animalium Angliae Tres Tractus. London; John Martyn (1678-81)

Said if fossils were organic, their originals were now extinct
RISE 262

MONTALBANO, MARCO ANTONIO DELLA FRATA ET (1635-1695)

Practica Minerale Trattato. Bologna; per Li Manolessi

Said all created beings had the power to multiply, but there was no spontaneous creation
AD 306

1679

BOCCONE, PAOLO (SILVIO) (1633-1704)

Museo di Fisica e di Esperienze Variato. Venezia; I.B. Zuccato (1697)

Said crude oils, bitumens and asphaltites differed only in consistency and that amber was a form of bitumin
FORB 29

KUNCKEL, JOHANN (1630-1703)

Ars Vitraria Experiminetalis. Amsterdam; Heinrich Betkio

Said minerals could be tested using a blowpipe
SMAJ 10

1680

BECHER, JOHANN JOACHIM (1635-82)

Chymisches Laboratorium, oder unter erdische Naturkundigung. Frankfurt; J. Haasz

Said sea water enters earth, heats, evaporates, rises through mountains and emerges as springs and rivers
BISWAS 181

LEIBNITZ, GOTTFRIED WILHELM VON (1646-1716)

Protogaea. Goettingae; Sumptibus I.G. Schmidii (1749)

Said fossils, but not crystals, were organic in origin
GEI 83-4

Said that as smooth, molten earth cooled, contracted, and formed a rough crust, inner condensation produced water, which filled ocean basins and arranged sediments into stratified layers and then receded into the earth as the crust alternately sank and rose, forming mountains and valleys
GEI 81-2; SCH 7; WOOD 16

Said the history of earth is shown in the successive layers of strata and the organic remains therein
GEI 83-4

Said that the early, molten, earth cooled and was covered by water under which the sedimentary strata formed until the lower layers collapsed
HMPD 335

Said igneous rocks were the result of fusion and aqueous rocks were the result of diluvial action which stratified sediments
WOOD 16-17

Said veins were filled either by liquifying fire or by water
TODH 312

NEWTON, ISAAC (1642-1727)

Letter to Burnet, in Turnbull: Correspondence of Isaac Newton, 2: 319 (1960). Cambridge; Published for the Royal Society at the University Press

Said earth's topography might be the result of uneven precipitation during primordial chaos
GLD 254

1681

BUCH, JOHANN JOACHIN

Physica Subterranea. Frankfurt; Imprimatur Mauritii Georgii Weidmanni

Said metals form from vapors rising from center of earth and more metals are formed deeper down
AD 286

1683

HATLEY, GRIFFITH

On Some Formed Stones of Fossil Found at Hunton in Kent, in Royal Society of London. Philosophical Transactions, 3: 4-5 (1683-84)

Said fossils grew in the soil
CAM 94

1684

BURNET, THOMAS (1635-1715)

Telluris Theoria Sacra (Sacred Theory of the Earth). London; R. Norton for W. Kettilby

Said early earth was a heterogeneous fluic then, a solid core formed within concentri density layers beneath smooth crust which, heated by the sun, cracked, forming mountains and valleys and releasing a flood of water which settled into the Pacific, Atlantic, and Mediterranean basir
MILL 157-8

Said mountains are destroyed externally by rain, wind, and sun´s heat, and internally by water, and eventually are leveled
GLD 258

Said the early earth had a hard core surrounded by concentric layers of water and earthy crust the outermost layer of which, hardened by the sun´s heat, ruptured, thus creating mountains and ocean basins and releasing the inner waters (the flood) most of which subsequently flowed back inside
CAM 98-9

Said the sun´s heat cracked the early earth´s crust releasing the oceans and creating mountains from the broken fragments
CHOR 11

Said the sun heated the equatorial region of the early earth, finally causing the waters to flow poleward where they condensed and rose creating precipitation, rivers and ocean currents
BISWAS 187-8

LANA, FRANCESCO TERZI (1631-87)

Magisterium Naturae et Artis, Opus Physico-Mathematicum. Brixiae; Io Mariam Ricciardum (1684-86)

Said the shape of crystals was determined by a saline principle during their formation by coagulation of water and nitrous material
AD 133; GORT 508

1684

MELZERN

Beschreibung der Stadt Schneebergt. Schneebergt; N.P.

Said warm vapors rising from ore deposits caused the snow to melt above mines
AD 303

1685

PLOT, ROBERT (1640-96)

De Origine Fontium Tentamen Philosophicum. Oxonii; Theatro Sheldoniano

Classified rivers as 1) temporary (regular and irregular), or 2) perennial (pure or mixed waters)
BISWAS 189

Said salt water rose to the peaks of mountains through a capillary system
BISWAS 190

1686

MARIOTTE, EDME (1620?-84)

Traité du Mouvement des Eaux et des Autres Corps Fluides. Paris; E. Michallet

Said rainwater did not penetrate soil very far, but flowed along impervious layers to form springs and rivers
BAKER 399; BISWAS 214-6

1687

BURNET, THOMAS (1635-1715)

Telluris Theoria Sacra (Sacred Theory of the Earth). London; Gualt Kettilby

Said 10,000 years would be sufficient to destroy earth´s present topography
GLD 259

HOOKE, ROBERT (1635-1703)

Discourse of Earthquakes, in R. Waller, Posthumous Works of Robert Hooke. London; S. Smith and B. Walford (1705)

Said spheroid shape of earth would cause axis and poles to move and alter the climate, cause volcanoes, and account for marine fossils in the land
STOKES 24-5

1688

GUGLIELMINI, GIOVANNI DOMENICO (1655-1710)

Riflessioni Filosofiche Dedotte dalle Figure de´Sali. Bologna; Antonio Pisarri

Said the planes and angles of any particular crystal were always inclined the same
GORT 508

1688

HOOKE, ROBERT (1635-1703)

Discourse of Earthquakes, in R. Waller, *Posthumous Works of Robert Hooke* (1705)

Said fossils could be used to date the strata
CONK 9; HED 23; OLD RHG 195; ZIT 19

Said that during the flood, the heavier land had sunk and the new lighter ocean floor rose creating the present fossiliferous strata
GLD 255

Said extinction of species and creation of new ones would occur oftenest during interchange of land and sea
ALB-2 50

1689

BARTHOLINI, THOMAS (CASPAR) (1655-1738)

De Fontium Fluviorumque in *Origine ex Pluviis*. Copenhagen; J.P. Bockenhoffer

Said that rainfall on any area was more than sufficient to maintain its rivers
AD 458

1690

HUYGENS, CHRISTIAN (1629-95)

Traité de la Lumière. Leide; Chez Pierre van der Aa

Said similar invisible flattened spheroids held side by side by air pressure, were cohesive and lent crystal axes their directional property
BJG 41-2, 67-8

WARREN, ERASMUS (d. 1718)

Geologia; or a Discourse Concerning the Earth Before the Deluge. London; R. Chiswell

Said creation took more than six days, the flood waters were not higher than the mountains and they did not alter the shape of the earth
CAM 46-58, 100

1691

HALLEY, EDMUND (1656-1742)

An Account of the Circulation of Watry Vapours of the Sea, and of the Cause of Springs, in Royal Society of London. Philosophical Transactions, 16: 468-73

Said a fluid outer core of the earth surrounded the solid innermost core
BRUSH 719

Said sea water was evaporated by the sun´s heat, rose and was carried to mountain heights where condensation turned it to rain and it fell, penetrated the earth and emerged in springs and rivers
BAKER 399; BISWAS 224-5

RAMAZZINI, BERNARDINO (1633-1714)

De Fontium Mutinensium Admiranda Scaturigine Tractatus Physio-Hydrostaticus. Modena; Typis Haeredum Suliani Impressorum Ducalium

Said artesian wells occur when waters are in constant motion
BISWAS 191

Said artesian wells were filled by seawater flowing through mountain reservoirs
AD 449-52

WHISTON, WILLIAM (1667-1752)

New Abstract of a Dissertation ... Concerning the System of the Earth, Its Duration and Stability. London; R. Roberts

Said a comet, alternately heated to liquid and cooled to freezing, was captured; then, specific gravity caused matter to sink, forming valleys of heavy above lighter mountain rock thus creating the earth
MILL 159-61; PORT 78

Said the deluge was caused by the condensation of vapor from a nearby comet´s tail plus waters which flowed out of the earth
MILL 159-61; WOOD 17

Said present earth was ordered and permanent in arrangement
PORT 75

Said the earth did not rotate on its axis until after the fall of man
WOOD 17

1692

RAY, JOHN (1627-1705)

Miscellaneous Discourses Concerning the Dissolution and Changes of the World. London; S. Smith

Said fossils had been deposited inland during a number of floods
CAM 101

Said the rivers and streams carry away and redeposit rain dissolved mountains, creating topography and new lands
CGT 2; CHOR 15; WOOD 39

Said fossils had once lived and each kind still had a living counterpart
CAM 101

1694

POMET, PIERRE (1658-99)

Histoire Générale des Drogues. Paris; J. B. Loyson

Said petroleum formed from the distillatio[n] of amber and fossil resins
FORB 30

1695

FONTANA, (CAJETANO) GAETANO (1645-1719)

Instituto Physico Astronomica. Modena; A. Capponi

Said anima of the geocosmos was responsibl[e] for movement of waters, rivers, and fountains
AD 442; OSR 9-10

LHUYD, EDWARD (1660-1709)

Letter to John Ray, August 28, in Life and Letters of Edward Lhuyd by Robert Gunther. Oxford; Subscribers (1945)

Said some fossils might be organic in orig[in]
RISE 260-1

WOODWARD, JOHN (1665-1728)

Essay toward a Natural History of the Earth and Terrestrial Bodies Especially Minerals. London; R. Wilkin

Said waters inside the earth fed springs, rivers, rains and oceans, and when it burst through the crust, bodies of plants and animals settled along with other debris to form fossiliferous strata, which internal heat dried and cracked into present topography
CAM 74-80, 104-7; GEI 67-8; SCH 9

1695

Said the earth was composed of rocks arranged in layers containing fossils
CONK 9

Said fossils were a product of the flood
PORT 74

Said ore veins were fissures which had filled with mineral and metallic matter drawn from the strata
HTOD 29

Said solid earth was dissolved by the flood and strata then was laid down according to the specific gravity of the suspended particles with fossils mixing in with the settling sediments
CGT 2; HMPD 334; PORT 77; RGIF 329

Said rocks were arranged in discrete layers which circled the earth
PORT 58-9

1697

GUGLIELMINI, GIOVANNI DOMENICO (1655-1710)

Della Natura de´Fiumi, Trattato Fisco-Mathematica. Bologna; A. Pisarri

Said less resistant stream beds produce less sloped channels and mobile beds are longitudinally concave upwards
CHOR 87

Said stream slopes would vary inversely with velocity or discharge
CHOR 87

Said streams erode or build up their beds, or their sides, until a balance is reached between force and resistance
CHOR 87

Said the velocity of water was proportional to the square root of the exit channel
BISWAS 193

SCHEUCHZER, JOHANN JACOB (1672-1733)

De Generatione Conchitarum, in Miscellanea Curiosa Medico-Physica. Academia Caesareo-Leopoldina Naturae Curiosorum, Series 3, Volume 4: Appendix 151-66

Said the laws of physics and chemistry explained the existence of fossils
SND 197

1698

LHUYD, EDWARD (1660-1709)

De Fossilium Marinorum et Foliorum Mineralium Origine (Letter to John Ray, July 29), in Lithophylacii Britannici Ichnographia Sive Lapidorum Aliorumque Fossilium Britannicorum Singulari Figura Insignium (Ground-Plan of the British Stone-Treasury). London; Ex Officina M. C. (1699)

Classified fossils as 1) lapides crystall 2) lapides corallini, 3) lithophyta, 4) fossilia turbinata, 5) bibalbia fossilia, 6) crustacea punctulata, 7) fossilia tubulosa, 8) malacostraca, 9) ichthyodontes cuspidati, 10) ichthyodontes scutellati, 11) xylostea, 12) ichthyospondyli, and 13) fossilia effigata anomala
COX 211; EPBG-1 144-5

Said that certain fossils occurred in different layers of strata
AD 258; WOOD 13

Said most fossils had grown from seeds deposited in the rock by rain or sea mist
EDW 6; HMPD 329; RISE 261; RUD 84; ZIT 17

1699

BECHER, JOHANN JOACHIM (1635-82)

Physica Subterraneae. Leipzig; J. L. Gleditoschium (1703)

Said natural materials were a combination of three earths, each of which conveyed certain properties: 1) vitrescible earth (fusibility or solidity, diaphaneity or opacity), giving rise to stones; 2) fatty earth (color, taste, odor, and inflammability), giving rise to earths; and 3) mercurial earth (metallic characteristics), giving rise to metals
MCR 56; SPMS 272-3

1700

LEIGH, CHARLES (1662-1701?)

Natural History of Lancashire, Cheshire, and the Peak in Derbyshire .... Oxford; C. Leigh

Said fossils were sports of nature and that crinoids were a natural mixture of sulfur, salts, and earth
EDW 5-6

LEMERY, NICOLAS (1645-1715)

Explication Physique et Chemique des Feux Souterrains, des Tremblemens de Terre, des Ouragans, des Eclairs et du Tonneur. Paris; L'Académie Royal des Sciences

Said fermentation in the earth produced earthquakes
AD 410

PRYME, ABRAHAM DE LA (1672-1704)

On the fossil shells and fishes in Lincolnshire, in Royal Society of London. Philosophical Transactions, 4: 521-4 and 22: 677-87

Said fossils were ubiquitous because flood was universal and were largely marine because land and ocean changed places
CAM 94; PORT 80-1

ROSLER, BALTHASAR (1605-73)

Speculum Metallurgia Politissimum oder, Hell-Polierter Berg-Bau-Spiegel. Dresden; Johann Jacob Wincklerin

Said veins were the result of fissures being filled
AD 310; TODH 314

STAHL, GEORGE ERNST (1660-1734)

De Ortu Venarum Metalliferarum (Short Discourse on the Origin of Metallic Veins). London; B. Bragg (1705)

Said mineral veins formed when the earth first formed and have changed little since then, though the Becherian earths (see Becher 1699) can produce metallic properties
SPS 270, 277

1702

HIARNE, URBAN (1641-1724)

Den Körta Anledningen, Til Atskillige Malm och Bergarters, Mineraliers och Jordeslags Etc. Eftersporjande och Angifwande, Beswarad och Forklarad. Stockholm; Första Flock on Watn

Said changes in sea level, deposition of eroded material, and surface deformation accounted for changes in coastlines
WEG 389

Said earth was constantly changing as mountains uplifted and wore away in an endless cycle
FRANG 346

Said exploding sulphur and other combustibles within the earth caused earthquakes and fiery eruptions
FRANG 346

Said flood disrupted the original order of the strata and accounted for location of some fossils
FRANG 346

Said the land was still rising
FRANG 346; FRANG-2 32

Said water circulated through channels in the earth drawn down by whirlpools in the ocean and routed up through mountains, emerging as springs and rivers
FRANG 346

SCHEUCHZER, JOHANN JACOB (1672-1733)

Specimen Lithographiae Helveticae Curiosae. Tiguri; Typis D. Gessneri

Said fossils were merely sports of nature
ZIT 20

TOURNEFORT, JOSEPH PITTON DE (1656-1708)

Description du Labyrinthe de Condie, avec Quelques Observations sur l´Accroissement et la Génération des Pierres, in L´Ami des Sciences: 217-34

Said mineral germs absorbed matter from the surrounding earth and then deposited the resulting particles on the walls of veins
BJG 21-22

1703

BAGLIVI, GEORGIO (1669-1707)

Opera Omnia. Lugduni; N.P. (1704)

Said growing minerals and stones are nourished by nutrients in solutions which enter their pores before solidification
AD 92

Said the moon caused an increase in earthquakes during winter when the earth is in perihelion
DAV 57

KONNIG, EMMANUEL (THE ELDER) (1658-1731)

Regnum Minerale. Basel; E. Konig Senioris

Said metals and other minerals were plants which were born and grew in the interior of the earth
AD 288

STAHL, GEORGE ERNST (1660-1734)

Comment, in Physica Subterranea (2nd Ed.) by J. J. Becher. Leipzig; J. L. Gleditschium

Said veins were filled by vapors rising from the earth´s interior
TODH 314

1705

GUGLIELMINI, GIOVANNI DOMENICO (1655-1710)

De Salibus Dissertatio Epistolaris Physico-Medico-Mechanica. Venetus; Aloysium Pavinum	Said the four basic minerals, 1) salt (cubic), 2) vitriol (rhombic), 3) niter (hexagonal), and 4) alum (octal) mixed with matter to form other crystals of the same form BJG 24-5

1708

LANG, CARL NICOLAUS (1670-1741)

Historia Lapidum Figuratorum Helvetiae. Venice; J. Tomasini	Said fossils formed inside rocks from dust borne germs of living species GEI 98; ZIT 18

SCHEUCHZER, JOHANN JACOB (1672-1733)

Bildnissen verschiedner Fischen (Notes upon the Plates in Piscium Querelae et Vindiciae). Zurich; Verlegung des Authoris	Classed fossils in the animal kingdom SND 199
Piscium Querelae et Vindiciae. Tiguri; Gessnerianis	Said fossils were remains of once living organisms AD 261

WOODWARD, JOHN (1665-1728)

Fossils, in Lexicon Technicum (by John Harris). London; D. Brown (1708-10)	Classified natural materials, based on exterior characteristics and composition, as 1) earths, 2) stones, 3) salts, 4) bitumens, 5) minerals, and 6) metals FRMS 192; PORT 55

1709

MYLIUS, GOTTLIEB FRIEDRICH (G.F.M.) (1675-1726)

Memorabilium Saxioniae Subterraneae. Leipzig; J. Groscheffen	Said fossils had been created by God and man would not be able to understand them AD 257

1710

BUETTNER, DAVID SIGISMUND (1660-1719)

Rudera Diluvii Testes. Leipzig; J. F. Braunen	Said fossils were deposited by the flood HMPD 330

1713

DERHAM, WILLIAM (1657-1735)

Physico-Theology. London; W. Innys

Said water rose to the tops of mountains through capillary action of the earth
BISWAS 256-7

RAY, JOHN (1627-1705)

Three Physico-Theological Discourses (Third Edition). London; W. and J. Innys

Said earthquakes and thunder were caused by fiery exhalations in the earth or sky
GEI 75

Said volcanoes occurred when earthquake fire encountered subterranean water
GEI 75

1714

MAZINI, GIOVANNI BATTISTA (1677-1743)

Congetture Fisico-Meccaniche Intorne le Figure delle Particello Componenti il Ferro. Brescia; G. M. Rizzardi

Said crystals formed from solutions and from molten minerals
GORT 508

RAY, JOHN (1627-1705)

The Wisdom of God Manifested in the Works of the Creation (Sixth Edition). London; William Innys

Said the sun evaporates seawater, wind drives it inland, and rising mountain vapors condense it to rain, which fills springs and rivers, and returns it to the sea
AD 259; BISWAS 186; GEI 74

1715

HOBBS, WILLIAM (1665?-1743)

Earth Generated and Anatomized. Ithaca; Cornell University Press (1981)

Said a day during creation was hundreds of times longer than it would be today
PORT-2 337

Said processes of creation each occurred only once and land and life occurred simultaneously
PORT-2 337

Said tides piled up layers of sand, and internal heat solidified the layers and sometimes caused them to uplift
PORT-2 336

1715

VALLISNIERI, ANTONIO (1661-1730)

Intorno all'Origine delle Fontane, in Address Delivered at University of Padua. Venice; Ertz

Said water from rain and melting snow created streams and, if trapped beneath impervious subsurface strata, would emerge as artesian springs
AD 452-8; BISWAS 254; OSR 14-5

1716-18

SCHEUCHZER, JOHANN JACOB (1672-1733)

Nature-Geschichte des Schweizerlandes (Natural History of the Swiss Landscape). Zurich; D. Gesner (1746)

Said fossils had been left by the flood
ZIT 20

1718

JUSSIEU, ANTOINE DE (1686-1758)

Histoire du Café, in Académie des Sciences, Paris. Mémoires, 31: 291

Said fossil plants had floated over France from warmer climates and been caught up during uplift
AD 261; RAM-2 24; ZIT 21

SWEDENBORG, EMANUEL (1688-1772)

Om Jordenes och Planeternas Gång och Stånd (On the Course and Position of the Earth and the Planets), in Opera, Volume 3. Holmiae; N.P. (1911)

Said the earth was losing orbital speed and receding from the sun, the soil was losing its fertility, and men their longevity, all of which suggested the gradual decay of the planet
FRANG 347-8

1719

STRACHEY, JOHN (1671-1743)

A Curious Description of the Strata Observed in the Coal-Mines of Mendip in Somersetshire, in Royal Society of London. Philosophical Transactions, 30: 968-73

Said rock structure was related to accompanying surface features
CHOR 17

Said shells in shale above a vein indicated there was coal below
CHOR 17; CONK 10; EGH 168

Said certain strata had an invariable order
EPGB-II 2

SWEDENBORG, EMANUEL (1688-1772)

Om Wattrens Högd och Förra Werldens Starka Ebb och Flood (On the Level of the Seas and the Great Fluctuations in Prehistoric Tides). Tryckeriet; J. H. Werner

Said erratics were worn smooth by the universal sea which tore them loose and then deposited them wherever they are now found
CGT 5; CHOR 197

Said fossils were remains of once living organisms
REG 173; STRO 4-5

Said mountains and hogbacks were formed from sedimentation under the ocean
FRANG 348

1720

MAILLET, BENOIT (1656-1738)

Telliamed ou Entretiens d´un Philosophe Indien avec un Missionaire Francois sur la diminution de la Mer, la Formation de la Terre, l´Origine de l´Homme, etc. Amsterdam; Chez l´Honoré et Fils Libraires (1748)

Said some fossil animals had no living counterpart because succession had changed them
CBS 216

Said wind, streams, glaciers, etc. had little effect on topography
ALB-2 70-7

Said all strata had been deposited sequentially over an uneven submarine surface
ALB-2 70-7; RHRE 85

Said earth was two thousand million years old, but man was only 400,000 years old
ALB-2 70-7; DEAN 447

Said life originated from the sea, fish giving rise to birds, and seaweed giving rise to shrubs and trees
ALB-2 70-7

Said oceans had once covered the earth, depositing fossil fish on mountains
ALB-2 70-7

Said rocks were deposited, piled up, and then permanently carved by the universal sea from which all life came, before the sea withdrew and the land emerged
CAR 140; UHS 386

Said heavenly bodies alternated as suns and planets and once water-covered earth is now evaporating again
CAR 140

REAMUR, RENE ANTOINE DE (1683-1757)

Remarques sur les Coquilles Fossiles de Quelques Cantons de Touraine, et sur les Utilités qu'on en Tire, in Académie des Sciences, Paris. Mémoires: 400-16 (1722)

Said inland marine fossils had been deposited by ancient arm of the sea a long time before the flood
CBS 215; DEAN 447

1721

RAY, JOHN (1627-1705)

Three Physico-Theological Discourses (Fourth Edition). London; W. and J. Innys

Said groups of fossils were communities which became fossilized when the waters withdrew from the land
CGT 2-3

VALLISNIERI, ANTONIO (1661-1730)

Dei Corpe Marini Che Su Monti Si Trovano. Venice; D. Lovisa

Said sea had once covered all the earth
AD 372; GEI 60-1

Said springs were located wherever the surrounding rock structure would permit water to rise up and out of the earth
GEI 60-1

1722

HENCKEL, JOHANN FRIEDRICH (1679-1744)

Flora Saturnizans. Leipzig; J. C. Martini

Said minerals formed during the original consolidation of the earth
SPMS 279

Said plant life could be understood by examining the growth process of minerals
SPMS 279

1723

CAPPELLER, MORIZ ANTON (1685-1769)

Prodromus Crystallographiae, de Crystallis Improprie Sic Dictis Commentarium. Lucernae; H.R. Wyssing

Classified forty stones, metals, earths, and salts into one of nine crystal types depending on the effect of origin on final shape
BJG 56-7; SMAJ 13

1723

SCHEUCHZER, JOHANN JACOB (1672-1733)

Itinera per Helvetiae Alpinas Regiones. Lugduni Batavorum; P. Vander Aa

Said water flowing down mountains entered cracks in glaciers, froze, and expanded, thus moving the glacier downhill
CGT 8; CHOR 192; GIA 158

1724

HALLEY, EDMUND (1656-1742)

Some Considerations about the Cause of the Universal Deluge, in Royal Society of London. Philosophical Transactions, 33 (383): 118-24

Said a comet struck earth, forcing the sea inland, and leaving mountains of heaped up sea bottom sediments when the sea receded
STOKES 26

REAMUR, RENE ANTOINE DE (1683-1757)

De l´Arrangement que Prennent les Parties des Matières Métalliques et Minérales, lorsqu´après avoir été Mises en Fusion, Elles Viennent à se Figer, in L´Ami des Sciences: 307-16

Said mineral crystallization was result of a crystalline essence or juice
BJG 27

1724

STUKELEY, WILLIAM (1687-1765)

Itinerarium Curiosum. London; Baker and Leigh (1776)

Said English strata tilted eastward in response to the earth´s rotation
EPBG-II 3

1725

HENCKEL, JOHANN FRIEDRICH (1679-1744)

Pyritologia, oder Kiess-Historie (History of the Pyrites). Leipzig; J. C. Martini

Said all substances fermented and gave off vapors which combine to produce metal deposits and minerals as they rise through the earth
AD 310-11; HTOD 31

Said all substances were formed by water plus earths (1) sulphur, the fat earth, 2) arsenic, the volatile earth, and/or 3) iron, the metallic earth
SPMS 282

Said crystals and stones formed from the union of earthy corpuscles cemented by calcareous earth mixed with water
SPMS 279

Said minerals might also grow by accretion
SPMS 283

Said mountain oils and pitches were organic in origin
FORB 34

Said only certain rocks had fissures suitable for vein formation
TODH 314

Said some minerals were continuously being created but that deep-seated ore bodies had been created during creation
SPMS 282

Said basalts were crystalline
AD 127

MARSIGLI, LUIGI FERDINANDO (1658-1730)

Histoire Physique de la Mer. Amsterdam; Aux Depens de la Compagnie

Said subsurface strata, classed as horizontal, oblique, or curved, continued for great distances, even beneath the ocean
RAM-2 25

Said that islands whose arrangement of strata was similar to that of mainland strata had been attached to that mainland
RAM-2 25

TATISCHCHEV, VASILII NIKITCH (1685-1750)

Manontova Cost, Hoc Est Ossa Subterranea, Fossilia, Ingentia, Ignoti Animalis e Siberia Adeferri Coepta, Durabus Perillustrium. Stockholm; N.P.

Said a heavy rain would have more effect the higher up a river encountered it
CHOR 434

Said that a river´s discharge would be proportionate to the size of the source area
CHOR 433

Said that a river´s rate of flow is determined by the size of the largest particles it carries
CHOR 430

Said that friction by water particles absorbed most of a river´s energy
CHOR 435

1725

Said the effects of rivers could only be determined by studying them in flood conditions
CHOR 432

Said water was non-elastic and therefore capable of eroding its channel
CHOR 433

Said rivers regulate themselves thus determining all their features
CHOR 430

Said mammoths were Siberian in origin and dated from a time when the climate there was much warmer
TDG 365

1727

BROMELL, MAGNUS VON (1679-1731)

Lithographia Svecana, in Acta Literaria et Scientiarum Sveciae, Volume 2, Trimestre Primum 63-77 and Trimestre Quartum 90-102 (1725-29)

Said fossils were organic remains left by the flood
PPHG 15

PRINCE, THOMAS (1687-1758)

Earthquakes, the Works of God and Tokens of His Just Displeasure. Boston; D. Henchman

Said earthquakes were caused by the explosive movement of vapors confined in the earth
HAZ 10

1728

PITOT, HENRI (1695-1771)

Principes d´Hydraulique. N.P.; N.P.

Said the resistance of a stream´s bed decreased the speed of the flow of its water
CHOR 87

WOODWARD, JOHN (1665-1728)

Attempt toward a Natural History of the Fossils of England. London; F. Fayram

Said some fossils were once living marine organisms
COX 211

1729

BOURGUET, LOUIS (1678-1742)

Lettres Philosophiques sur la Formation des Sels et des Crystaux et sur la Generation et le Méchanisme Organique des Plantes et des Animaux. Amsterdam; Francois l'Honoré

Said erratics had fallen from the sky
GIA 158

1730

NEWTON, ISAAC (1642-1727)

Opticks. London; W. Innys

Said crystals formed when tiny particles in an evaporating liquid were forced together by mutual attraction
MILL 519-20

1731

POLHEM, CHRISTOPHER (1661-1751)

Epistola ad Andream Celsium, in Acta Literaria et Scientiarum Sveciae, Volume 3 (1730-34). Uppsala; Gottfrid Kiesewetter

Said the earth was several thousand years old
FRANG 347

Said the interior of the earth was 10,000 kilometers across, full of fire, and surrounded by 500 kilometers of thick pumice beneath the outer crust
FRANG 347; REG 172-3

1732

PLUCHE, NOEL ANTOINE (1688-1761)

Spectacle de la Nature. London; J. Pemberton (1733)

Said ocean waters rose through sand and earth to mountain tops to emerge as springs and rivers
BISWAS 257

1734

COLONNE, MARIE-POMPEE (1644-1726)

Histoire Naturelle de l'Univers. Paris; A. Cailleau

Said sand was a salt with much earth in it and was produced by sea water along the shore
AD 128

Said mountains grew like plants in response to the heat of the earth
AD 363

HENCKEL, JOHANN FRIEDRICH (1679-1744)

Idea Generalis de Lapidum Origine. Dresden and Leipzig; N.P.

Said pebbles are no longer forming but were created when the soft, spongy early earth hardened
SPMS 280

1735

LINNAEUS, CARL VON (1707-78)

Oratio de Telluris Habitabilis Incremento. Lugduni Batavorum; Theodorum Haak (1743)

Said earth was originally water-covered; then an island formed, and the original species evolved there
WEG 391

Systema Naturae. Lugduni Batavorum; Theodorum Haak (1743)

Said species retain their definitive characteristics through time and would be found living in the sea nearest where they were found in fossil form
PPHG 36-7; WOOD 20

Classified minerals according to their crystalline shapes (external and internal) into species, genera, and classes
FRMS 192-3; MILL 146; SCH 67; SMAJ 14; WOOD 20

Said fossils were minerals
HMPD 338

Classified earth materials as 1) stones (refractory, calcareous, and vitrescible), 2) minerals (salts, sulfur, and mercurials or 3) fossils (earths, concretions, and petrifactions)
OLD NS 508

Said crystallization was result of saline action with each salt affecting specific materials to form a given shape
BJG 58-9; MCR 66

1736

PACKE, CHRISTOPHER (1686-1749)

A New Philosophico-Chorographical Chart of East Kent. London; J. Roberts (1737)

Said dry valleys were the result of the flood
GEI 451

1738

TILAS, DANIEL (1712-72)

En Bergsmans Rön och Försök i Mineral Riket (On a Mining Expert's Experience in Mineralogy). Turku; I.C. Merckell

Said ore could be found among erratic blocks since blocks were larger and commoner near the origin of both
FRANG 32-3

1739

RICHMANN, GEORG WILHELM (1711-53)

On Remarkable Changes to Which the Surface of the Earth is Subjected from Time to Time, in Primechaniia na Vedomosti, Part 89-96

Said changes in the landscape were due to gravity-driven water carrying mountain tops down into depressions
TDG 365

1740

BOUGUER, PIERRE (1698-1758)

La Figure de la Terre. Paris; C. A. Jombert (1749)

Said that the gravitational attraction of the Andes Mountains was less than their mass would indicate it should be
DALY 1

MORO, ANTON LAZZARO (1687-1740)

De Crostacei. Venice; S. Monti

Classified mountains as 1) primary, raised by magma (unstratified), 2) secondary (sedimentary, stratified), and 3) primary core with a secondary outer crust
GEI 61-5; HED 24; OLD RHG 196

Said early earth was covered by fresh water until inner fires fractured the crust and elevated it above the water which then received sediments and salt from the eroding land onto which marine plants and animals moved
GEI 61-5

Said heated gases within the earth expand, creating cavities and uplifting the strata above, after which surrounding strata subside and fill the cavity as gases retreat
BEL 25

Said volcanoes were instrumental in mountain and island building
AD 365-72

1741

STOBAEUS, KILIAN (1690-1742)

Monumenta Diluvii Universalis ex Historia Naturali. London; L. Decreaux

Said fossils are remains of life forms which perished in the flood
PPHG 2

1742

BOURGUET, LOUIS (1678-1742)

Traité des Pétrifications. Paris; Briasson

Said whirlpools of primitive matter formed the sun and planets, settling according to specific gravity of constituents thus forming earth´s crust and strata
AD 129

Said four hundred forty-one figures demonstrated the succession of life forms but assigned no particular age to them
CBS 215

CAT, CLAUDE NICHOLAS LE (1700-1768)

Traité des Sens. Rouen, Paris; G. Cavelier

Said early earth was muddy sphere and the effect of the sun and moon caused land to rise as the sea excavated the ocean basins
MILL 163-4

TILAS, DANIEL (1712-72)

Stenrikets Historia, Utförd i Det Wid Praesidii Afläggande Håldne Talet in för Kongeliche Swenska Wetenskaps - Academien den 14 April 1742, in Kongeliche Wetenskaps Academien Handlingar, 1: 194-201. Stockholm; L. L. Grefing

Said fossils are petrifications of life forms
PPHG 3

1743

EVANS, LEWIS (1700-56)

His Journal, in Pownall, Thomas: Topographical Description of Such Parts of North America as Are Contained in the Map of the Middle British Colonies in North America. London; J. Almon

Said rains weathered high lands and carried them to the sea thus allowing the lightened land to rise
LEEA 302

Said fossils were carried up with mountains when land rebounded following the receding of the oceans into their basins
CHOR 236-8

Said fossils were once living marine organisms
LECT 152

Said the Appalachian Mountains were remains of a plain which existed prior to the flood and was cut into valleys by running water
CHOR 236-8; LECT 125

TILAS, DANIEL (1712-72)

Mineral-Historia Ofver Osmunds-Berget Uti Rattwiks Sochn och Oster-Dalarne, in Kongeliche Vetenskaps Academiens Handlingar, 1: 194-201

Said erratics were deposited by icebergs afloat in a universal sea
CGT 7; CHOR 197

1744

BUFFON, GEORGES LOUIS LECLERC (1707-88)

Histoire et Théorie de la Terre, in Histoire Naturelle (1749-67). Paris; Imprimerie Royale

Said strata formed in water
RGIF 331

Histoire Naturelle. Paris; L'Imprimerie Royale (1749-1803)

Said planets were chunks detached from the sun by a comet
GEI 89; P&S 10

Said strata could be dated by the fossils deposited in them by the universal ocean
GEI 89, 190

Histoire Naturelle (Second Printing?). Paris; L'Imprimerie Royale (1750)

Said earthquakes were 1) felt for long distances and not accompanied by volcanic activity, or 2) local tremors caused by pyrites igniting underground as a volcano heats up
DAV 5, 8

Théorie de la Terre, in Histoire Naturelle (1749)

Said earth was very old and had developed over 75,000 years
D&B 32; Dean 447; PPHG 7

Said rain waters carried the land into th ocean where currents arranged the debris into mountains and valleys which were revealed when the ocean once again over-flowed the now leveled land
HMPD 334

Said there were no general rules deter-mining the arrangement of the earth´s strata
OLD RHG 228

CELSIUS, ANDERS (1701-44)

Oration de Mutationibus Quae in Superficie Corporum Coelestium Contingunt. Lugduni Batavorum; Cornelium Haak

Said sea level was falling steadily at a rate of approximately four feet per century
WEG 390

HENCKEL, JOHANN FRIEDRICH (1679-1744)

Introduction à la Minéralogie. Dresden; J. E. Stephanus (1747)

Classified earth materials as 1) earths (refractory and fusible), 2) stones, 3) salts, 4) earthy juices and sulfurous substances, or 5) metals
OLD NS 508

Classified stones by appearance and their reaction to acid as 1) calcareous, 2) siliceous, 3) calcareous-siliceous, and 4) argillaceous
SPMS 281

MARTEL, PETER

An Account of a Journey to the Glacieres or Ice Caps in Savoy. London; for Peter Martel

Said movement of glaciers was a result of the gravitation of stones and sand down valleys
CHOR 192, 295

1746

GUETTARD, JEAN ETIENNE (1715-86)

Mémoire et Carte Minéralogique. Académie des Sciences, Paris. Mémoires: 363-92

Said French formations disappeared under the English Channel and reappeared in England so France and England were part of the same geological region
ALB-3 225; GEI 110-5

POTT, JOHANN HEINRICH (1692-1777)

Lithogeognosia. Berlin; C. F. Voss (1757)

Classified earth materials as earths and stones which were 1) alkaline and calcareous, 2) gypseous, 3) argillaceous, or 4) siliceous
OLD NS 508; SPMS 284

Classified minerals according to chemical reaction to heat
MILL 47

Said it is impossible to determine the interior composition of substances
SPMS 285

ZIMMERMAN, CARL FRIEDERICH (1713-1747)

Obersachsische Bergakademie in welcher die Bergwerkswissenschaften nach ihren Grundwahrheiten untersucht und entworfen werden. Dresden und Leipzig; N.P.

Said veins were parent rock altered by a mineralizing saline solution and transformed into metals
AD 311; HTOD 31; TODH 314-5

1747

WALLERIUS, JOHAN GOTTSCHALK (1709-1785)

Mineralogia Eller Mineral-Riket (Systema Mineralogicum). Stockholm; L. Salvii

Classified natural materials by appearance as 1) earths (dry, argillaceous, mineral, or sandy), 2) stones, sulfurs, semimetals, or metals), and 4) concretions (fossils and pseudofossils)
FRANG-2 35-6; HTOD 134; LIND 95; OLD NS 508; PPHG 47-9

Said all matter was a combination of three basic earths (vitrescible, sulphurous, mercurial) and their corresponding mineral acids
SPMS 286-7

Said mineral crystals form from earthy and metallic substances, the latter of which have a unique shape which is imposed on the crystal
BJG 26; LIND 103

Said fossils were remains of living organisms which died in the flood
PPHG 2, 48

1748

WOLTERSDORFF, JOHANN LUKAS (D. 1772)

Systema Minerale in Quo Regni Mineralis Producta Omnia Systematice per Classes, Ordines, Genera et Species Proponuntur. Berlin; C. L. Kunst

Classified earth materials as 1) earths (argillaceous or alkaline), 2) stones (alkaline, gypseous, argillaceous, or vitrified), 3) salts, 4) bitumens, 5) semimetals, 6) metals, or 7) petrifications
OLD NS 508

1749

BUFFON, GEORGES LOUIS LECLERC (1707-1788)

Histoire Naturelle. Paris; De l´Imprimerie Royale (1749-1803)

Said species are constant and do not generate new, different species
D&B 62

Histoire Naturelle des Minéraux, in Histoire Naturelle.

Said decomposition of organic matter combined with feldspar, basalt, water, etc. to create crystals
BJG 23

Said metals grow much like plants with organic molecules on the outer edges determining form
AD 296; MSSC 154

Théorie de la Terre, in Histoire Naturelle (1749)

Said as molten earth cooled, topography formed under the oceans, and later uplifte[d] as now new topography is still forming under present oceans
HOOY 276-7

Said 1) a comet detached a piece of the sun, 2) the fireball cooled, solidifed an[d] crinkled into mountains, 3) then the ocean covered what is now dry land, 4) the earth ruptured, admitting ocean waters, 5) the crust split into old and new worlds, and 6) erosion of the land began
CHOR 13-14; P&S 10

GENERELLI, GIUSEPPE CIRILLO

Dissertazione de´Crostacei, e dell´altre Produzioni Marine che Sono ne Monti, in Raccolta Milanese dell´Anno 1757. Milano; Nella Stamperia di Antonio Agnelli (1757)

Said constant degradation of mountains by running water caused land to rise
GEI 126-7

OPPEL, FRIEDRICH WILHELM VON (1720-69)

Anleitung zur Markacheidekunst nach ihren Anfangsgründen und Ausubung kürtzlich entworfen (Introduction to Subterranean Geometry). Dresden; G. C. Walther

Said bedded deposits occurred conformably but vein materials had entered from outside the country rock
AD 311

Said that veins cross the structure of country rock and often parallel other veins
AD 311; TODH 315-6

Said veins were result of drying and cracking of country rock or movements of the earth´s crust and subsequent infilling of the cracks by mineral-rich waters
AD 311

1750

BOSCOVICH, RUDJER JOSIP (1711-87)

De Letteraria Expeditione per Pontificiam Ditonem, in De Bononiensi Scientiarum et Artium Instituto atque Academia Commentarii, 4: 353-96 (1757)

Said the Andes´ low density mass at depth compensated for the mass of the mountains
MAB 61

STUKELEY, WILLIAM (1687-1765)

On the Causes of Earthquakes, in Royal Society of London. Philosophical Transactions, 46: 641-6

Said exhalations rose from the earth´s interior, condensing into metallic veins, and releasing electricity which caused 1) earthquakes and 2) lightning
AD 411, 414

1751

LEHMANN, JOHANN GOTTLOB (1719-67)

Kurtze Einleitung in Einige Theile der Bergwerckswissenschaft. Berlin, Leipzig; Christoph Gottlieb Nikolai

Said the sun was instrumental in forming gold, so it was found in hot climates, while cooler metals were found in cooler climes
AD 282-3

1751-52

TARGIONI-TOZZETTI, GIOVANNI (1712-83)

Voyage Minéralogique, Philosophique, et Historique, en Toscane (Travels in Tuscany). Paris; Lavilette (1792)

Said resistance of rocks controls only the rate at which running water can erode
CHOR 16; MATH 74

Said valleys were cut by rivers
CHOR 16

1752

GUETTARD, JEAN ETIENNE (1715-86)

Mémoire sur Quelques Montagnes de la France Qui Ont été des Volcans (Memoir on Certain Mountains in France Which Have Been Volcanoes), in Académie des Sciences, Paris. Mémoires: 27-59 (1756)

Said that volcanoes occur when mountains containing coal, bitumen, etc., catch fire
FENT 44

Said the earth's internal fires once fed now extinct volcanoes and still heat hot springs
GEI 34

Said the Auvergne Mountains were extinct volcanoes
GEI 27-31

1753

LEHMANN, JOHANN GOTTLOB (1719-1767)

Abhandlung von den Metall-Muttern und der Erzeugung der Metalle. Berlin; G. Nicolai

Said metals formed 1) in the depths of the earth growing upward, like the limbs of a tree, or 2) in pockets of the crust from vapors rising from the "limbs"
AD 286, 292; TODH 315

PEYSONNEL, CLAUDE CHARLES DE (1727-90)

Traité du Corail, in Royal Society of London. Philosophical Transactions, 47: 445-69

Said corals were animals
AD 132; STODD 199

1753-54

BRAHMS, ALBERT

Anfangsgründe der Deich- und Wasserbaukunst. Aurich; H. Tapper

Said channel friction was proportional to the area of the stream cross-section divided by the wetted perimeter
CHOR 87

Said downstream velocity suggested equal and opposite effect of channel friction as opposed to gravity imposed acceleration
CHOR 87

1754

KALM, PEHR (1716-79)

Exercitium Academicum de Ortu Petrificatorum. N.P.; N.P.

Said fossils were petrified animals and plants which have been scattered over the earth by floods and earthquakes
PPHG 3-4

ROUELLE, GUILLAUME FRANCOIS (1703-70)

Cours de Chemie. Clifton College (Manuscript)

Said there were six earths or stones: 1) calcareous (shell remains), 2) gypsum, 3) argillaceous, 4) siliceous, 5) fusible, and 6) refractory, some of which are interchangeable, and all of which derive from an earthy principle
SPMS 296

1755

BOSCOVICH, RUDJER JOSIP (1711-87)

Philosophiae Naturalis Theoria (Theory of Natural Philosophy). Venetia; A. Bernardi

Said particles attract or repel other particles at certain locations on their surfaces thus arranging themselves in a predetermined manner
BJG 49-50

CARTHEUSER, FRIEDRICH AUGUST (1734-96)

Elementa Mineralogiae, Systematice Disposita. Frankfort; N.P.

Classified earth materials as 1) earths, 2) stones (both 1 & 2 were further divided based on appearance), 3) salts, 4) inflammables, 5) semi-metals, or 7) heteromorphs
OLD NS 509

D´ARGENVILLE, ANTOINE JOSEPH DEZALLIER (1680-1765)

L´Histoire Naturelle Eclaircie dans une de ses Parties Principales, l´Oryctologie qui Traite des Terres, des Pierres, des Métaux, des Minéraux et Autres Fossiles. Paris; De Bure L´Aine

Classified earth materials as 1) earths, 2) stones (both 1 and 2 further divided based on appearance), 3) salts, 4) sulfurs, and 5) metals
OLD NS 509

GELLERT, CHRISTLIEB EHREGOTT (1713-95)

Anfangsgründe zur metallurgischen Chymie (Metallurgic Chymistry). London; for T. Beckett (1776)

Classified earth materials as 1) earths (argillaceous or alkaline), 2) stones (calcareous, gypseous, argillaceous or vitrifiable), 3) salts, 4) sulfurs, 5) semi-metals, or 6) metals
OLD NS 509

KANT, IMMANUEL (1724-1804)

Allgemeine Naturgeschichte und Theorie des Himmels (Universal Natural History and Theory of the Heavens and Mechanical Origin of the Whole Universe). Königsberg; Johann Friederich Petersen

Said that a primeval rotating cloud of gas, a nebula, condensed to form several balls which are now planets
P&S 8

WINTHROP, JOHN (1714-79)

Lecture on Earthquakes. Boston; Edes and Gill

Said the movement of gases through passage in the earth caused progressive waves which moved faster than the speed of sound and culminated in an earthquake, thus releasing subterranean pressures
FGA 1829; HAZ 11

1756

C. E. F.

Neue Sammlung merckwürdiger Geschichte. Breslau and Leipzig; Pietsch

Said gold spread like vegetables grow
AD 296

LEHMANN, JOHANN GOTTLOB (1719-1767)

Traité de la Formation des Métaux, in Traite de Physique, d'Histoire Naturelle, de Mineralogie et de Métallurgie.

Said germs of metals are held in matrices (ore bodies, veins, fissures) until released into mixes of the proper proportions to form the appropriate metal (incorrect proportions result in semi-metals)
SPMS 294-5

Said metals formed from a combination of three earths [vitrifiable, phlogiston (subtle earth plus saline particles), and mercurial] and which metal formed was determined by the latter two ingredients
SPMS 293-4

Traité de Physique, d'Histoire Naturelle, de Minéralogie et de Métallurgie. Paris; J. T. Herissant (1759)	Said the earth's crust was a series of layers, each different, laid one upon another on an uneven submarine surface AD 217; RHRE 85, 88
Versuch einer Geschichte von Flötzgebürgen. Berlin; F. A. Lange	Said earthquakes and volcanoes were essential to the development of the earth's topography VON 23
	Described the stratigraphic succession in thirty layers of rock CONK 10
	Said mountains had a tripartite structure: 1) primitive, fossil-free, mineral rich core, formed during creation, 2) floetz-gebirge, surrounding primitive ranges, stratified, formed during the flood; and then 3) alluvium, formed from local sediments AD 374-8; ALB-2 117; CHOR 18; HED 25; OLD RHG 198

LILIENTHAL, THEODOR CHRISTOPH (1717-82)

Die gute Sache der göttlichen Offenbarung. Königsberg; Hartung (1760-82)	Said the flood split the crust of the earth into the present continents CDBN 1-2

1757

COSTA, EMANUEL MENDES DA (1717-91)

A Natural History of Fossils. London; L. Davis and C. Reymers	Classified earths and stones as fossils but ignored true fossils WOOD 23

JUSTI, JOHANN HEINRICH GOTTLOB VON (1720-71)

Grundriss des gesamten Mineral-reiches Worinnen alle Fossilien Göttingen; Wittwe	Classified earth materials as 1) earths and stones (precious, semi-precious, calcareous, refractory, or fusible), 2) salts, 3) combustibles, 4) semi-metals, 5) metals, or 6) petrifactions OLD NS 509

1757-59

LOMONOSOV, MIKHAIL VASILEVICH (1711-65)

O Sloyakh Zemnyh (On the Strata of the Earth), in First Principles of Metallurgy or Mining. St. Petersburg; Imperial Academy of Sciences (1763)

Said topography was result of 1) external forces (wind, rain, rivers, fires) and 2) internal forces (earthquakes-uplift and subsidence)
BEL 28

Said mountains emerged from the sea after being laid down horizontally, then were inclined by the force of the earth´s inner heat
BEL 29; UHS 387

1758

CRONSTEDT, AXEL FREDRIK (1722-1765)

Essai d´une Nouvelle Minéralogie (Essay towards a System of Mineralogy). Stockholm; J.A. Carlbohm

Classified earth materials as 1) calcareou 2) siliceous, 3) garnet kind, 4) argillaceous, 5) micaceous, 6) fluores, 7) asbest kind, 8) zeolites, 9) manganese kind, 10) salts, 11) inflammables, 12) semi-metals, 13) metals, or 14) petrifactions
OLD NS 509

Said chemical tests divided rocks and minerals: minerals were 1) earth, 2) salt 3) inflammables, and 4) metals, while rocks were 1) rocks (compound), 2) petrifaction, and 3) slags
FAIR 152; HTOD 134; LIND 103; MILL 518

ISNARD, DE GRASSE

Mémoire sur les Tremblemens de Terre. Paris; Vve David Jeune

Said mineral exhalation of metals and ores included an electric fluid which caused earthquakes
AD 411

MACQUER, PIERRE JOSEPH (1718-84)

Dictionnaire de Chymie, Contenant la Théorie et la Pratique de Cette Science. Paris; Lacombe (1766)

Said there were individual earths but most contained a terrene principle unifyin various combinations of impurities
MCR 55-6, 67

Elements of the Theory and Practice of Chemistry. A. Millar and J. Nourse

Said petroleum and bitumen originated from chemical rocks composed of plant oils and acids
FORB 35; OWEN 15

1758-59

LINNAEUS, CARL VON (1707-1778)

Systema Naturae (Tenth Edition). Holmiae; L. Salvii	Classified earth materials as 1) rocks, 2) ores (salts, sulfurs, and metals), and 3) fossils (soils, agglomerations and petrifications) BJG 58; HTOD 133-4; PPHG 33

1759

ARDUINO, GIOVANNI (1714-95)

Saggio Fisico-Mineralogico di Lythogonia e Orognosia (and Letters to Antonio Vallisnieri) in Atti dell´ Accademia delle Scienze di Siena: 76 (1780?) and in Nuovo Raccolta di Opuscoli Scientifici e Filogici del Padre Abate Angiolo Calogiera, 6: 99-180	Recognized four divisions of earth´s crust: 1) primary primitive mountains containing minerals, 2) secondary mountains of limestone and marble, 3) tertiary low mountains and hills of gravel, sand, and marine debris, 4) quaternary alluvial materials of 1, 2, and 3 washed from the mountains by streams AD 373-4; CONK 10
Osservazione Sulla Fisica Constituzione delle Alpi Venete, in Raccolta di Memorie Chimico-Mineralogiche, Mettalurgiche, e Orittografiche (1775)	Differentiated between sedimentary and igneous rocks and described contact metamorphism GORT 511

LOMONOSOV, MIKHAIL VASILEVICH (1711-65)

Oratio de Generatione Metallorum (Pervye Osnovaniia Metallurgii ili Rudnykh Del (Erste Grundlagen der Metallurgie oder des Hutten-wesens) (Generation of Metals from the Earth´s Quaking). St. Petersburg; N.P. (1763)	Said earth´s changing topography was the result of very slow natural activity (erosion, uplift, shifting oceans, etc.) TIK 366; VON 23

176?

ROUELLE, GUILLAUME FRANCOIS (1703-1770)

Notice sur sa Doctrine Relative à Plusieurs Points Importans de l´Histoire Naturelle de la Terre, in Encyclopédie Méthodique: Géographie Physique by Desmarest. Paris; H. Agasse (1795)	Said changes in the polar axis had caused climatic changes OLD RHG 229 Said rocks were 1) ancient or 2) new (containing nonrandom fossil groups which varied from place to place and could be used to name the bed since some were always together and some never were) OLD RHG 229

176?

Said there had been alternations in the distribution of land and sea
OLD RHG 229

1760

DESMAREST, NICHOLAS (1725-1815)

Observations Géologiques, in Oeuvres de Mr. Turgot, 3: 376-447 (1808-11)

Said small-scale erosion by running water in time would remove beds of strata
TND 344

GRUNER, GOTTLIEB SIGMUND (1717-78)

Die Eisgebirge des Schweizerlandes. Bern; A. Wagner

Said glaciers move by sliding over the ground which is lubricated by their melting edges
GIA 158

LOMONOSOV, MIKHAIL VASILEVICH (1711-65)

Discourses on the Solidity and Liquidity of Bodies, in Complete Collection of the Works of Lomonosov. Moscow; Academy of Science SSR (1952)

Said crystal shapes were determined by the arrangement of atoms
TDG 368

MICHELL, JOHN (1724-93)

Conjectures Concerning the Cause and Observations upon the Phenomena of Earthquakes; Particularly of That Great Earthquake of the First of November 1755, Which Proved So Fatal to the City of Lisbon and Whose Effects Were Felt as Far as Africa and More or Less Throughout Almost All of Europe, in Royal Society of London. Philosophical Transactions, 51 (2): 566-634

Said knowledge of local fossils was necessary in order to trace strata
CONK 10

Said strata had formed horizontally, then bent up in the middle to form mountains and leave strips of minerals along the rises
GEI 379-80; PORT 120-1

Said volcanoes and earthquakes were involved in uplift as well as in the fracturing and rearranging of strata
RGIF 332

Said an earthquake was waves of vibration which moved at uniform speeds within an ar[ea] but varied from place to place and was followed by aftershocks
DAVIS 15-20; DAV 68

Said exact location of a marine epicenter could be determined because the waves would move at the same speed
DAV 22

Said epicenter could be located by noting times of arrival of shock waves and then drawing interconnecting lines
DAV 22

Essay on the Causes and Phenomena of Earthquakes, in Royal Society of London. Philosophical Transactions, 49 (1761)

Said earthquake motions were partly tremulous and, further away, wavelike
GEI 274-6; WOOD 24

Said earthquakes were caused by sudden addition of large quantities of water to subterranean and volcanic fires
AD 415-8; GEI 274-6; WOOD 24

Said epicenter was focus of strongest waves which radiated and diminished outward so lines could be drawn to locate it
GEI 274-6; WOOD 24

Said volcanoes may be caused by the spontaneous combustion of pyritous strata
AD 415-8; GEI 174-6; WOOD 24

1761

CATCOTT, ALEXANDER (1725-79)

A Treatise on the Deluge. London; M. Withers and D. Prince

Said much erosion was the result of the universal flood
WOOD 39

Said continuity and horizontality of strata dated from creation but folded strata, unconformities, and detritus were caused by the flood
NEVE 48, 51-2; WOOD 39

Said fossils were deposited when flood destroyed and redeposited original strata
NEVE 51-2

FUCHSEL, GEORG CHRISTIAN (1722-73)

Historia Terrae et Maris ex Historia Thuringiae per Montium Descriptionem Eruta, in Acta Academiae Electoralis Morguntivae zu Erfurt, 2: 44-209

Divided regional rocks into lithologic formations representing one continuous strata per epoch of time, i.e., of the environment of formation
GEI 199; SCH 10

Said each stratum was laid down in one period of time and each period could be denoted by its particular layer
GEI 199; VON 23; OLD RHG 200

Said formations were continuous strata of the same composition laid down in one epoch
GEI 199; VON 23

Said when earthquakes fracture rocks, surface waters then fill the spaces to form veins
GEI 199

Said fossils might indicate the age of the strata in which they are found
AD 269

MICHELL, JOHN (1724-93)

Conjectures Concerning the Cause and Observations upon the Phenomena of Earthquakes; Particularly of That Great Earthquake of the First of November 1755, Which Proved So Fatal to the City of Lisbon and Whose Effects Were Felt as Far as Africa and More or Less Throughout Almost All of Europe, in Royal Society of London. Philosophical Transactions, 51 (2): 566-634

Said the shock wave had moved at more than twenty miles per minute
AD 415-8

ROBINET, JEAN BAPTISTE RENE (1735-1820)

De la Nature. Amsterdam; Chez E. von Harrevelt

Said fossil germs produce minerals, stones and metals, like kind from like kind
BJG 22

1762

DESMAREST, NICHOLAS (1725-1815)

Remarques de Mr. Desmarest sur la Géographie Physique, les Productions et les Manufactures de la Généralité de Bordeaux, in Archives Departmentales de la Dordogne (ms. 26)

Said that running water had carved the valleys and continued to do so
FENT 48; TND 345

1763

BAUME, ANTOINE (1728-1804)

Manuel de Chymie. Paris; Chez La Combe (1766)	Said all matter forms from combinations of earth, water, fire and air SPMS 299

DESMAREST, NICHOLAS (1725-1815)

Encyclopédie, ou Dictionnaire Raisonné des Sciences, des Arts et des Métiers. Paris; Briasson, David Lebreton (1768)	Said that basalt was volcanic TND 347-8

LOMONOSOV, MIKHAIL VASILEVICH (1711-65)

Erste Grundlagen der Metallurgie oder des Hüttenwesens (Generation of Metals from the Earth's Quaking). St. Petersburg; N.P.	Said metals formed as a result of fire and heated gases in the earth BEL 30

RASPE, RUDOLPH ERICH (1737-94)

Specimen Historiae Naturalis Globi Terraquaei. Leipzig; J. Schreuder and P. Martin	Said folding and fracture of the sedimentary layers was result of earthquakes causing uplift RHRE 89
	Said mountains have a core of uplifted fossil-free basement rock covered by a sedimentary layer RHRE 88
	Said the earth solidified beneath the oceans where earthquakes and subterranean explosions broke the crust and eventually pushed the debris-covered beds above the surface to form islands, continents, and mountains MILL 168; RHRE 89-91
	Said vertical uplift caused tilting and folding of beds, accompanied by fracturing which resulted in veins infilled by subterranean solutions RHRE 88

1764

WALCH, JOHANN ERNST IMMANUEL (1725-78)

Systematic Description of the Mineral Kingdom, in Das Steinreich. Halle; J. J. Gebauer (1762-64)

Said basalt columns were composed of cryst
AD 127

1765

GUETTARD, JEAN ETIENNE (1715-86)

On the Accidents That Have Befallen Fossil Shells Compared to Those Which Are Found to Happen to Shells Now Living in the Sea, in Académie des Sciences, Paris. Transactions: 189, 329, 399

Said fossil beds found inland corresponded to those found in the oceans
CONK 10

Said fossils were remains of once living organisms
FENT 42

1766

Bergman, Torbern Olof (1735-84)

Physisk Beskrifning öfver Jordklotet (A Physical Description of the Globe). Uppsala; N.P.

Classified rocks as 1) oldest, nonfossiliferous, interior of mountain, 2) bedded, shells encircling the earth, result of erosion, some fossils, 3) youngest, swept together in shallow waters, and 4) also young, volcanic
REG 177-9; HED 27, 34, 37; TBS 189

Said a change in the inclination of the ecliptic would change climates
PPHG 5

Said continuity of strata from one side of a valley to the other suggested the crust had been torn apart
HED 29

Said fossils found inland had been deposited by floods
PPHG 5

Said fossils had been placed on primeval crust and were old as the world
TBS 188

Said mountains were formed by both chemical and mechanical processes and ran primarily from north to south because of the tides
FRANG 351; REG 177-9; HED 39

1766

Said basalt formed when it was softened by volcanic vapor and then cracked by drying
REG 177-9

BRANDER, GUSTAVIUS (1720-87)

Fossilia Hantoniensia. London; W. Wood (1829)

Said many fossils were of warmer water species than those which lived nearby
WOOD 23

1766-68

LINNAEUS, CARL VON (1707-78)

Systema Naturae (Twelfth Edition). Holmiae; L. Salvii

Named and classified minerals according to observable surface features (of the ´unfolded´ crystal)
FRMS 193

1769

HAMILTON, WILLIAM (1730-1803)

Letter Containing Some Farther Particulars on Mount Vesuvius, in Royal Society of London. Philosophical Transactions, 59: 18-22

Said volcanoes were natural force for construction, not just destruction, and had raised mountains and enlarged coastal areas
PORT 161-2

RASPE, RUDOLF ERICH (1737-94)

A Letter Containing a Short Account of Some Basalt Hills in Hassia, in Royal Society of London. Philosophical Transactions, 61: 580-3 (1771)

Said basalt was igneous in origin
BEER 107; PORT 175

1770

DELIUS, CHRISTOPH TEAUGOTT (1728-79)

Abhandlung vom Ursprung der Gebirge und der darinne befindlichen Erzadern. Leipzig; Bey C. G. Hilschern

Said veins were fissures caused by drying and contraction of rocks which were then filled by minerals carried into the rock by water circulating through the earth
HTOD 31-2; TODH 315

Said veins decreased at depth when sun´s rays no longer could affect them
AD 311-3

GUETTARD, JEAN ETIENNE (1715-86)

Mémoires sur Différentes Parties des Sciences et des Arts. Paris; L. Prault

Said the ocean was the most important eroder of shorelines but ineffective in deep water
GEI 125

Said debris of erosion ended up on shores of the ocean unless deposited along stream beds
AD 261; GEI 122

Said mountains were lowered by rain, rivers and the ocean
CHOR 17

Said surface erosion, by rain and rivers, was very fast
CHOR 17

On the Basalt of the Ancients and the Moderns, in Mémoires sur Différentes Parties des Sciences et des Arts, Volume 2. Paris; L. Prault

Said basalt was a vitrifiable rock which crystallized in an aqueous fluid
GEI 135

LAVOISIER, ANTOINE LAURENT (1743-94)

Premier (and Second) Mémoire sur la Nature de l'Eau, et sur les Expériences par Lesquelles on a Prétendu Prouver la Possibilité de Son Changement en Terre, in Académie des Sciences, Paris. Mémoires: 73-82, 90-107 (1773)

Said water maintained its identity and could not turn into earth
AD 124

TOALDO, GIUSEPPE (1791-98)

Della Vera Influenza Degli Astri. Padova; G. Manfre

Said the winter moon (in perihelion) might have caused an increase in the number of earthquakes
DAV 57

1771

BOSSUT, CHARLES (1730-1814)

Traité Elémentaire de Hydrodynamique. Paris; C. A. Jombert

Said rivers of like depth, width, slope, and bed material have the same speed and if one factor changed, the speed would change in a predictable manner
CHOR 88

HAMILTON, WILLIAM (1730-1803)

Remarks upon the Nature of the Soil of Naples and Its Neighborhood, in Royal Society of London. Philosophical Transactions, 61: 1-43

Said change in nature was very slow and uniform
PORT 161, 233; SLEEP 325

RASPE, RUDOLF ERICH (1737-94)

A Letter Containing a Short Account of Some Basalt Hills in Hassia, in Royal Society of London. Philosophical Transactions, 61: 580-3

Said basalts were volcanic in origin
PORT 175

1772

HAMILTON, WILLIAM (1730-1803)

Observations on Mount Vesuvius, Mount Etna and Other Volcanoes. London; T. Cadell

Said pumice developed from bitumin
SLEEP 331-2

KERN, JOHANN GOTTLEIB (1745?-75)

Bericht von Bergbau. Leipzig; S. L. Crusius

Said vapors from the interior where minerals grow, rise and warm the earth and air above
AD 304

L'ISLE, JEAN BAPTISTE LOUIS DE ROME DE (1736-90)

Essai de Cristallographie. Paris; Chez Didot Jeune

Classified minerals according to external characteristics (geometry) with consideration of shape, specific gravity, and hardness
FRMS 194

Said crystal forms were determined by a saline principle
BJG 63-4; MCR 66

Said crystals were 1) saline, soluble in water, 2) stony, resistant to fire, 3) sulfurous or arsenious, fumes in fire, or 4) metalllic, fluid in fire
BJG 63-4

Said gems differed from rocks by being composed of hard platelets stacked on top of each other
PHIL 13

1772

Said given crystals formed from the combining of identical molecules and final shape would be determined by purity of solution, formation time, and the amount of agitation that occurred during that time
BJG 63-4

1773

BAUME, ANTOINE (1728-1804)

Chymie Expérimentale et Raisonée. Paris; P. F. Didot le Jeune

Said marine animals capture fire (sunligh during their production of calcareous earth, thus making it available for the formation of salts, metals, etc.
SPMS 301

BERGMAN, TORBERN OLOF (1735-84)

Physical and Chemical Essays. London; J. Murray (1784-91)

Said successively reducing the layers of various crystal faces accounted for the various shapes in which a substance might form
SMAJ 12

BORDIER, ANDRE CESAR

Voyage Pitoresque aux Glacières de Savoye. Genève; L. A. Caille

Said that a glacier was a heavy, viscous mass and moved accordingly
DGGS 74

1773-74

BERGMAN, TORBERN OLOF (1735-84)

Physisk Beskrifning öfver Jordklotet (A Physical Description of the Globe) (Second Edition). Uppsala; Comographiska Sällskapet

Said as universal water evaporated, botto sediment compacted and crystallization occurred in cavities as topographic highs caused by uneven settling and currents were exposed to air and life began thereo
HED 36

Said boulders found on mountains were carried there encased in floating ice
HED 40

Said chemical reaction could change a shell to calcite, silica or an ore
HED 38

Said early earth was fluid so equator bulges and poles have been flattened by rotation
HED 35

Said fossils were hard animal and plant parts whose soft organic matter decomposed slowly in a wet, solid matrix which permitted solutions to enter the hard parts
HED 31, 38; PPHG 4

Said granite had a sedimentary origin
HED 37

Said rocks are composed of salts, earths, burnable materials and metals
HED 35

Said basalt formed after its substance was softened by volcanic gases and then dried
FRANG 351; REG 177-9

Said character of uppermost rocks determined the steepness of a mountain slope
HED 29

1774

WERNER, ABRAHAM GOTTLOB (1749-1817)

Von den äusserlichen Kennzeichen der Fossilien (On the External Characters of Minerals). Leipzig; S. L. Crusius

Classified minerals primarily by external and physical features and place of origin, and used composition for fossil identification
AD 201-4; BJG 59-61; MILL 149; OLD RHG 201; SMAJ 14

Said true classification of minerals could be achieved only by chemical analysis
GNC 33

1775

ARDUINO, GIOVANNI (1714-95)

Raccolta di Memorie Chimico-Mineralogiche, Metallurgiche, e Orittografiche. Venice; B. Milocco

Said parts of the earth had repeatedly uplifted and/or subsided as a result of injections of magma
AD 374

Said trap rocks were volcanic in origin
WOOD 19

CHEZY, ANTOINE DE (1718-98)

Mémoire sur la Vitesse de l´Eau Conduite dans une Regolé, in Annales des Ponts et Chaussées, 60 (1921)

Calculated speed of river water to be Chezy´s coefficient times hydraulic radius times slope, the product raised to the 1/2 power
CHOR 88, 100

CRONSTEDT, AXEL FREDRIK (1722-65)

Versuch einer Mineralogie. Gratz; Brunnich

Classified minerals according to effect of fire, oil, water: 1) unaffected, 2) dissolve in oil and burn in fire, 3) dissolve in water, and 4) fluid in fire
WMC 65-6

GRIGNON, PIERRE CLEMENT (1723-94)

Mémoires de Physique sur l´Art de Fabriquer le Fer, d´en Fondre & Forger des Canons d´Artillerie. Paris; Delalain

Said fusion of metals was caused by fire
BJG 29

Said crystals were the result of the slow cooling of molten material
BJG 29

STRANGE, JOHN (1732-99)

An Account of ... Giants Causeway, in Royal Society of London. Philosophical Transactions, 65: 5-47, 418-23

Said basalts were part of lava flows and other volcanics
PORT 174

1776

HAMILTON, WILLIAM (1730-1803)

Campi Phlegraei. Naples; P. Fabris

Said basalt was volcanic in origin
SLEEP 324

Said earthquake intensity diminished outward from a center around which an isoseismic map could be drawn
SLEEP 328

Said the earth was many thousand years old and changed slowly and uniformly
SLEEP 324-5

Said the earth´s inner heat fermented minerals whose exhalations caused earthquakes with lightning, and upwelling lava encountering water, caused volcanoes
SLEEP 328-30

Said the earth's inner heat uplifted land, islands and mountains, as well as causing volcanoes and earthquakes
PORT 161-2; SLEEP 325, 328

Said volcanoes and earthquakes were release valves for magmatic pressures in the earth
SLEEP 328

WALLERIUS, JOHAN GOTTSCHALK (1709-85)

Meditationes Physiochemical de Origine Mundi, Imprimis Geocosmi Ejusdemque Metamorphosi (Origin of the World). Stockholm and Upsaliae; M. Swederus (1779)

Said creation was a mixing of an earthy and a fire principle with varying results affected by affinity (coalescence) and gravity (unequal drying) and the interconvertibility of water and earth
SPMS 289-92

Tänkar om Verldenes, I Synnerhet Jordenes, Danande och Ändring (Thoughts on the Creation of the World). Stockholm; H. Fougt

Said God created light (sun and stars) and water (earth) which then gave rise to all other substances
FRANG 349-50; FRANG-2 36

1777

BERGMAN, TORBERN OLOF (1735-84)

Disquisitio Chemica de Terra Gemmarum, in Regiae Societas Scientiarum Upsaliensis Nova Acta, 3: 137-70

Said chemical analysis of minerals involved definite sequences of reactions
MCR 57; WMC 67

Said minerals could best be analyzed by humid methods
FRMS 196

Said there were six elementary earths:
1) siliceous, 2) calcareous,
3) argillaceous, 4) magnesian,
5) ponderous, and 6) diamond
SPMS 303

PALLAS, PETER SIMON (1741-1811)

Observations sur la Formation des Montagnes et les Changemens Arrivés au Globe, Particulièrement à l'Egard de l'Empire Russe, in Academia Scientiarum Imperialis Petropolitana Acta: 21-64

Said mountains had been uplifted
CHOR 20

Said that when the earth solidified, granite mountain peaks appeared as islands which waves eroded, thus creating crystalline schist slopes, subsequently covered by fossiliferous calcareous secondary rocks, and finally by tertiary clays and marls
AD 378; CHOR 20

Werner, Abraham Gottlob (1749-1817)

Kurze Klassifikation und Beschreibung der verschiedenen Gebirgsarten (Short Classification and Description on the Different Rocks). Dresden; in der Watherischen Hofbuchhandlung (1787)

Said basalt was aqueous, not volcanic, in origin
GEI 222

Said between primitive rocks being exposed by receding ocean and uplifting to form mountains, and erosion-formed floetz, marine life arose with terrestrial fossils appearing in later floetz rocks
CHR 25-6; HAZ 222-3

Said deposits could be dated by their fossils
WOOD 34

Said fossils were deposited from universa oceans on several occasions (different fossils at different times)
GEI 215

Said granite was found on mountains as well as in valleys because universal ocean deposited it everywhere onto an uneven core
OSPO 166

Said rocks of primitive mountains appeare in a chronological sequence: granite, gneiss, micaceous schist, argillaceous schist, primitive limestone, trap, porphy syenite, serpentine, topaz rock, quartz, and siliceous schist
OLD RHG 201

Said the composition of rocks was not random and could be determined
WKK 246

1777-79

LINNAEUS, CARL VON (1707-78)

Vollständiges Natursystem des Mineralreiches. Nurnberg; G. N. Raspe

Said ocean sand crystals precipitated out of seawater
AD 128

1778

BEROLDINGEN, FRANZ COELESTIN VON (1740-98)

Beobachtungen, Zweifel und Fragen die Mineralogie überhaupt, und insbesondere ein naturliches Mineral-System betreffend. Hanover; J. W. Schmidt

Said coal and petroleum were formed by the transformation of organic matter
OWEN 10-5

BUFFON, GEORGES LOUIS LECLERC (1707-78)

Epoques de la Nature, in Histoire Naturelle (Natural History). Paris; L´Imprimerie Royale

Said life appeared after 35,000 years, the continents separated after 65,000 years, and the earth was 75,000 years old
ALB-2 85

Said life originated at north pole, which cooled first thus permitting formation of organic molecules from oily water and ductile earth in a succession of unrelated species which migrated southward as cooling continued
ALB-2 83

Said creation took 85,000 years
DEAN 452

Said original topography formed on the sea floor where currents swept sediments into mountains and valleys
ALB-2 82

Said wood of primitive forests, died, was buried, caught fire and created volcanoes
ALB-2 82

Said calcareous rocks are composed of shellfish fragments
ALB-2 81

Said Atlantic Ocean was formed by Atlantis sinking and the American continent being eaten away by westerly currents
CDBN 1

Said world is constantly changing and its temperature is falling
HOOY 277-8

Said early earth was water covered, water receded into interior, land appeared and topography began to form
WOOD 21; SCH 8

Said earth formed in six steps: 1) a sta cools, rounds, and consolidates;
2) topography appears and metal veins forn
3) waters cover earth, deposit fossils on mountains, recede, and forests appear;
4) receding waters and vegetation sink into fissures, combine with metals and he and create volcanoes; 5) animal life appears, climate is warm throughout northern hemisphere; 6) land mass separate into new and old world
GEI 91-5; RGIF 331; SCH 7

Said new organisms would appear, disappear, be replaced by new sequential but unrelated organisms, and so the cycle would go on
ALB-2 83

CHARPENTIER, JOHANN FRIEDRICH WILHELM VON (TOISSANT VON) (1738-1805)

Mineralogische Geographie der chursächsischen Lande. Leipzig; S. L. Crusius

Said veins and other ore bodies were result of alteration of parent rock
AD 313-4; POGS 129

FAUJAS DE ST. FOND, BARTHELEMY (1742-1819)

Recherches sur les Volcans Eteints du Vivarais et du Velay. Genoble; J. Cuchet

Said columnar basalt was formed by the contraction of the earth´s crust when it cooled
HSP 53

FORSTER, JOHANN REINHOLD (1729-98)

Theory of the Formation of Isles, in Observations Made During a Voyage Round the World, on Physical Geography, Natural History, and Ethic Philosophy. London; G. Robinson

Classified Pacific islands as 1) high and volcanic, or 2) low and coralline
STODD 199

Said corals built in a circle to create a sheltered refuge
STODD 199

LUC, JEAN ANDRE DE (1727-1817)

Lettres Physiques et Morales sur l´Histoire de la Terre, et de l´Homme. La Haye; De Tune (1779)

Said basalts were of volcanic origin
TND 353

Said that earth was completely covered by ocean, then mountains rose to form islands, then continents, and life flourished until volcanic activity destroyed the crust and water again covered the globe, thus beginning a new cycle
MILL 171-2

Said volcanic activity and weathering had more actively eroded the land early in earth´s history
PORT 165

Lettres Physiques et Morales sur les Montagnes. En Suisse; Les Libraires Associés

Said interior of earth was filled with air which was forced out through fissures whenever subsidence of land occurred, thus rearranging and fracturing the ocean floor
CGT 7

Said erratics were rocks ejected from inner earth by volcanoes
CHOR 198; GIA 158

LUDOVICUS

Pyrotechnia Sublimis. Vienna; N.P.

Said metals mature when exposed to heat and eventually turn into gold
AD 297

MORVEAU, LOUIS BERNARD, GUYTON DE (1737-1816)

Elemens de Chymie: Théorique et Pratique, Volumes 1-2. Dijon; L. N. Frantin

Said crystals formed from cooling solutions of salts dissolved in water or from metals dissolved in fire
BJG 29

PRYCE, WILLIAM (1725?-90)

Mineralogia Cornubiensis. London; J. Phillips

Said veins which cross other veins reveal their relative ages
HTOD 46

Said water absorbs particles as it moves through the earth and later precipitates minerals and metals in fissures to form veins
HTOD 32

WHITEHURST, JOHN (1713-88)

An Inquiry into the Original State and Formation of the Earth. London; J. Cooper

Said coal formed from vegetable matter
EPBG-II 15

Said earth was very old
PORT 166

Said some lava flowed between strata and hardened after the formation had been formed, actually burning some of the original strata
CBS 216; EPBG-II 14-5

Said the entire universe was once fluid and the earth, by rotating, formed layers around a spheroid with a universal ocean where life originated, islands formed and sediments accumulated and were uplifted by internal heat to form continents
MILL 169-70; PORT 125

Said there was a regular order to strata and their inclusions
CONK 13; EPBG-II 13; GEI 380

1779

BERGMAN, TORBERN OLOF (1735-84)

Commentario de Tubo Ferrumintatio. Vindobonae; I. P. Kraus

Said minerals could be tested by use of a blowpipe
SMAJ 10

Opuscula Physica et Chemica. Uppsala; Officinis Librariis Magni Swederi (1779-90)

Classified minerals as 1) salts, 2) earth 3) metals, 4) phlogistica, 5) petrae (mechanical products), and 6) petrafacta (organic fossils)
MCR 56

Classified minerals by analytical chemical methods
SMAJ 10

Said there were five basic earths (ponderous, calcareous, argillaceous, and siliceous) which, in combinations of two, three, four, and five, produced 325 natural materials
FRMS 196-7; MCR 56

BLUMENBACH, JOHANN FRIEDRICH (1752-1840)

Handbuch der Naturgeschichte. Göttingen; J. C. Dieterich

Devised a time scale based on fossils
CBS 216

BUAT, PIERRE LOUIS GEORGES DU (LOUIS GABRIEL) (1737-1809)

Principes de Hydraulique, Vérifiés par un Grand Nombre d'Expériences Faites par Ordre du Gouvernement. Paris; L'Imprimerie de Monsieur

Said rivers cut their valleys and wore the highlands down with the help of rainwash
CHOR 88-92

Said rivers flowed faster over large stones than over gravel and meandered to achieve grade
CHOR 88-92

Said rivers have four ages: infancy, youth, middle course, and old age
CHOR 88-90, 92

BUFFON, GEORGES LOUIS LECLERC (1707-88)

Epoques de la Nature, in Histoire Naturelle. Paris; L'Imprimerie Royale (1749-1803)

Said the earth was 78,000 years old
CBS 216

LUC, JEAN ANDRE DE (1727-1817)

Lettres Physiques et Morales sur l'Histoire de la Terre, et de l'Homme. La Haye; De Tune

Said erratics in southern Jura were alpine in origin and had been moved by catastrophic catapulting
BEER 107

MONNET, ANTOINE GRIMOALD (1734-1817)

Nouveau Système de Minéralogie. Bouillon; à la Société Typographique

Classified minerals in accordance with the four earths which, in combination, made up all other natural materials
MCR 59

SAUSSURE, HORACE BENEDICT DE (1740-99)

Voyages dans les Alpes. Neuchâtel; S. Fauche

Said granite mountains crystallized out of the primeval sea
AD 388

Said strata were laid down under water in successive layers
CHOR 30

1779

Said erratics were transported by violent currents when Alps were uplifted
GIA 158

Said inclined strata had been deposited on an incline
UHS 392

Said strata were laid down horizontally underwater by successive depositions, but might later be folded by some great force
AD 389-90

Said valley cutting was done by snow and rain water
MARG 100

Said the extent of the advances and retreats of glaciers could be determined by the location and composition of moraines
CHOR 198

Said earthquakes split the crust and the oceans drained inward, uncovering the mountains and valleys
CHOR 33

Said erosion was done by rivers and rainwater
WOOD 25

Said erratics and large valleys were result of a violent flood of water
CHOR 31, 198

Said glaciers moved when subterranean heat melted the ice freezing them to the ground
GIA 158

Said glaciers were moved by gravity
DGGS 74

Said primitive mountains had been uplifted above their valleys to form the Alps
AD 392

Said sub-glacial water was melted by earth's inner heat
CHOR 30

Said there might be succession in the appearance of some fossils
WOOD 25

WERNER, ABRAHAM GOTTLOB (1749-1817)

Notes on Geognosie. (Manuscript only?)

Said the core of the earth was metallic and deposits were offshoots from the core
OSPO 165-6

1780

FICHTEL, JOHANN EHRENREICH VON (1732-95)

Beytrag zur Mineralgeschichte von Siebenburgen. Nurnberg; in Verlag der Raspischen Buchhandlung

Said that salt and petroleum had evaporated from the sea and were trapped in underground cavities which formed when the earth´s crust solidified
OWEN 17

Mineralogische Bemerkungen von den Karpathen. Wien; J. von Kurzbeck (1791)

Said volcanic activity produced 1) broad intrusive masses, or 2) surficial lava flows
OWEN 16

SOULAVIE, JEAN LOUIS GIRAUD (1752-1813)

Histoire Naturelle de la France Méridionale (Natural History of Southern France). Paris; J. F. Quillau (1780-84)

Said limestone could be age related by its fossil forms: 1) primordial (extinct forms), 2) mixed (extinct and now living forms), 3) fossils of living forms, 4) fossils plus vegetation, and 5) a mixture of conglomerate and modern alluvium plus fossils
CBS 217; D&B 24, 26; DEAN 448-9

TOULMIN, GEORGE HOGGART (1754-?)

Antiquity and Duration of the World. London; T. Cadell

Said earth is infinitely old and man also is everlasting, but both are continually destroyed and rebuilt
GHT 346-50

Said internal heat of earth causes earthquakes and volcanoes
GHT 340

Said near destruction of earth and life occurred in the past (and will again)
DEAN 452

Said tomorrow´s continents are developing on today´s ocean floors
GHT 340

1781

DOLOMIEU, (DEODAT) GRATET DE (1750-1801)

Voyage aux Iles de Lipari (Travels in the Lipari Isles). Paris; Rue et Hotel Serpente (1783)

Said fluidity of magma was combination of heat plus some components, perhaps sulfur, which resulted in mobility and scoriaceous texture of the solidified rock
HTOD 41

GERHARD, CARL ABRAHAM (1738-1821)

Versuch einer Geschichte des Mineralreichs. Berlin; N.P.

Said movements of the earth opened fissure which circulating waters then filled with dissolved minerals creating veins
AD 315; TODH 315

1782

FRANKLIN, BENJAMIN (1706-90)

Conjectures Concerning the Formations of the Earth, in American Philosophical Society Transactions, 3: 1-5 (1793)

Said center of earth was gaseous
NCDAFG 245

Said as earth formed, escaping energy produced heat, and gravity drew heavy particles down to melt and form metal-rich-fluid core, create the magnetic field, fuel volcanoes, and cause magnetic north to wander and the planet to rotate
FGA 1827; HAZ 189

HAUY, RENE JUST (1743-1822)

Extrait d´un Mémoire sur la Structure des Spaths Calcaires, in Journal de Physique, 20: 33+

Said there was one primal form for all calcites and that variations were result of wear and tear on edges and corners
KUNZ 67-8

1783

CAROSI, JAN FILIP DE (1744-1801)

Sur la Génération du Silex et du Quartz en Partie. Cracovie; I. Grebel

Said limestone changed into quartzose flint
BAIL 88

L´ISLE, JEAN BAPTISTE LOUIS DE ROME DE (1736-90)

Cristallographie, ou Description des Formes Propres à Tous les Corps du Règne Minéral. Paris; L´Imprimerie de Monsieur

Said all crystals were derived from six basic primitive forms
SMAJ 14

Said crystal faces of a given species are invariable in their angles
BJG 69; SMAJ 14; TIK 326

VIVENZIO, GIOVANNI

Istoria e Teoria de'Tremuoti in Generale; ed in Particolare di Quelli della Calabria e di Messina del 1783. Naples; Nella Stamperia Regale

Said earthquakes were caused by electricity
AD 420

1783-84

HALL, JAMES (1761-1832)

Tours Abroad, 9 May 1783 to 14 April 1784. Edinburgh; National Library of Scotland (Manuscript)

Said successive volcanic episodes created new land
PORT 162

1783-88

BUFFON, GEORGES LOUIS LECLERC (1707-88)

Histoire Naturelle des Minéraux, in Histoire Naturelle (Natural History). Paris; Imprimerie Royale

Said early molten earth cooled into five primitive glasses (quartz, jasper, feldspar, schorl, and mica) which subsequently formed various mineral substances
FRMS 199

1784

BERGMAN, TORBERN OLOF (1735-84)

Manuel du Minéralogiste, ou Sciagraphie du Règne Minéral. Paris; Chez Cuchet

Said bitumens were derived from decomposition of vegetable and animal products by action of mineral acids
OWEN 15

Meditationes de Systemate Fossilium Naturali. Florentiae; Typis J. Tofani Prostan and Venales apud a Carlieneri

Classified inorganic material according to its chemical composition
LIND 135

Said there were four classes (salts, earth, metals, and phlogistin) of natural material which combined to form other substances
FRMS 197

DAUBENTON, LOUIS JEAN MARIE (1716-99)

Tableau Méthodique des Minéraux, Suivant Leurs Differentes Natures, et avec des Caractères Distinctifs, Apparens ou Faciles à Reconnoître. Paris; Chez Villie	Classified minerals by their external characteristics MCR 61; WMC 67

HAUY, RENE JUST (1743-1822)

Essai d'une Théorie sur la Structure des Crystaux Appliquées à Plusieurs Genres de Substances Crystallisées. Paris; Chez Gougue et Née de la Rochelle	Said elementary molecules of various chemicals combined to form constituent molecules of various shapes BJG 108
	Said crystals were result of stacking together tiny, identical integral molecules DANA 3

KIRWAN, RICHARD (1733-1812)

Elements of Mineralogy. London; P. Elmsley	Classified all earth-related bodies according to qualitative chemical criteria excluding objects of vegetable or animal life CIR 42; MCR 59; WMC 67
	Said most mountain water evaporated before it could carry much soil down, much less reach the ocean CHOR 50

L'ISLE, JEAN BAPTISTE LOUIS DE ROME DE (1736-90)

Des Caractères Extérieurs des Minéraux. Paris; L'Auteur	Said there was a one-to-one relationship between the external appearance of minerals and their composition and physical properties FRMS 194-5

PIGNATORO, DOMENICO (1735-1802)

Giornale Tremuotico, in Descrizione de Tremuoti Accaduti nelle Calabrie nel 1793. Naples; N.P. (1788)	Constructed an earthquake intensity scale: 1) slight, 2) moderate, 3) strong, 4) very strong, and 5) violent AD 420-1; DAV 28

785

OUGLAS, JAMES (1753-1819)

Dissertation on the Antiquity of the Earth. London; Logographic Press	Said climatic cycles recorded in fossils and strata proved they had formed over a long period of time PORT 159-60

utton, James (1726-97)

Abstract of a Dissertation ... Concerning the System of the Earth, Its Duration and Stability, in Royal Society of Edinburgh. Transactions, 1: 209-304 (1788)	Said earth had no apparent beginning or ending DEAN 454; GNC 20; PORT 189
	Said earth´s central fire caused volcanoes, earthquakes, and mineral veins PORT 188
	Said fossil wood and coal was a remnant of earlier plant life HUBB 228; FENT 65
	Said earth had a cycle: 1) older rock becomes sediment on the seafloor, 2) is consolidated by heat and pressure, 3) is heated to expansion and fractured, 4) molten material flows upward into fractures driven by volcanic action, 5) emerging above sea level, the land begins to wear away again AD 386-7; ALB-2 94-5, 229; DON 7; MILL 173; MOR 93-4; PORT 187-8
	Said new rock formed continuously on the seafloor from river-carried debris of eroded highlands RAM-1 35
	Said present crust was result of ancient sedimentation collecting on the seafloor, subsiding, consolidating due to heat and pressure, fracturing, and being uplifted by volcanic action or earthquakes AD 240; CHOR 42-3; CORB 27; MILL 173; WOOD 35-42
	Said sedimentary rock formed on the ocean floor from mechanical and chemical processes, but intrusives were volcanic in origin and sediments deposited on molten lava would be consolidated by the heat and intruded by the lava ALB-2 93-5; CHOR 43-4

1785

Said strata formed horizontally, then bed
were tilted by force from earth´s hot
interior while still buried in the earth
AD 240; GNC 20

Said subterranean heat consolidated
sediments deposited on ocean floor, causi
expansion and uplift, with fracturing and
subsequent inflow of igneous rock and
mineral solutions
AD 240, 386-67; CHOR 43-4

Said that geological time was endless
HUBB 225-55; BAD 159; BAIL 50; CORB 27-8;
D&B 35

Said that some species of marine life
had always existed but that man was of
recent origin
HUBB 228; GHT 345

Said that when stresses of pressure build
up in the interior, they are released
through volcanoes, or by earthquakes
AD 386-7; BAIL 44-5; GEI 304

Said there is an ageless, repetitious,
uniformity in evolution of the earth and
present day processes are the keys to
past history
CHOR 37; CORB 28; GEI 291; PORT 194;
RAN 321-2

Said unconformities indicate an interruption in the deposition of strata
CHOR 37; CORB 27; GNC 20; UHS 395-402

Said weathering and valley-cutting rivers
slowly but surely changed the topography
although uplift was relatively fast
HUBB 229; BAIL 50; CORB 27-8; D&B 35

Said that all rock species, excluding
some minerals, could be found in England
WOOD 35-42

Said unconformities were the key to identi
fying the cycles of earth´s development
CONK 12

Said the earth functions like a heat machi
ALB 2

MODEER, ADOLPH (1739-99)

Anledning Til Stenrikets Upställning På Stadgade Grunder, Second Edition. Stockholm; N.P.

Said fossils should be classified in the same way as living organisms
PPHG 49

SOULAVIE, JEAN LOUIS GIRAUD (1752-1813)

Les Classes Naturelles des Minéraux et les Epoques de la Nature Correspondante à Chaque Classe. St. Petersburg; L'Académie Impériale des Sciences

Classified minerals by event rather than form so that class stood for age, order stood for year, species stood for phenomenon, and variety stood for occurrence
OLD RHG 234-5

TREBRA, FRIEDRICH WILHELM HEINRICH VON (1740-1819)

Erfahrung von Inneren der Gebirge. Leipzig; Gelehrte und Kunstler

Said ore deposits formed from a process of fermentation and rotting; i.e., heat and humidity converted constituents of some rocks into metals
AD 315; TODH 315

1786

LEBLANC, NICHOLAS (1742-1806)

Essai sur Quelques Phénomènes Relatifs à la Cristallisation des Sels Neutres, in Journal de Physique, 28: 341-45 (see also his De la Cristallotechnie ou Essai sur les Phénomènes de la Cristallisation, in Journal de Physique, 45: 296-313 (1802)

Said crystals formed in one of two shapes depending on 1) chemical composition, 2) proportions of constituents, and 3) the physical conditions of the original solution
BJG 114-5

SAUSSURE, HORACE BENEDICT DE (1740-99)

Voyages dans les Alpes, Volume 2. Neuchâtel; S. Fauche

Said some strata were laid down horizontally but strata forming under water might form on an incline
MARG 100

Said strata were laid down horizontally under water and later inclined
UHS 392

Said erosion was the result of running water
GREG 105

WERNER, ABRAHAM GOTTLOB (1749-1817)

Kurze Klassifikation und Beschreibung der verschiedenen Gebirgsarten (Short Classification and description on the different rocks), in Kongeliche Böhmische Gesellschaft der Wissenschaften Abhandlungen: 272-97. Dresden; In Der Watherischen Hofbuchhandlung (1787)

Said fluid earth contained all rock elements and became stratified as heavier materials sank to form granite core, etc., absorbing water, leaving present ocean waters and forming topography by settling, erosion, and uplift
CHOR 25; GEI 220-2, 307; OSPD 165-6

Said inclusions were older than the surrounding rock
SEDD 388

Said there were four categories of rocks, in sequence, worldwide: 1) primitive, oldest crystalline (chemically formed: granite, gneiss, schist, etc.), 2) floetz, stratified and overlying the primitive rocks (chemically formed with some mechanical action: limestone, sandstone, etc.), 3) volcanic, true (partially chemical, but largely mechanical: surficial lava, pumice, ash, and tuff) and pseudo (lavalike layers melted by internal burning of coal beds), and 4) alluvial (mechanically worked 1) and 2), and sometimes 3) with gradations between each
ALB-2 120-2; CHOR 25-6; GEI 214-5; GNC 41; HAZ 222-3

1787

GADD, PEHR ADRIAN (1727-1797)

Ron Och Undersokning, I Hwad Man Insecter Och Zoophyter Bidraga Til Stenhardningar Stockholm; Kongeliche Vetenskaps Academien Handlingar

Said earth had been created many thousand years ago
PPHG 7

HAIDINGER, KARL (1756-97)

Systematische Eintheilung der Gebirgsarten. Wien; C. F. Wappler

Said rocks were either 1) primary or 2) secondary
HTOD 136

LASIUS, GEORG OTTO SIGISMUND

Beobachtungen ueber die Hartzgebirge. Hannover; in Der Helwingischen Hofbuchhandlung (1789)

Said materials in water circulating through the earth were captured by carbonic acid and solvents and deposited in veins
TODH 315

WERNER, ABRAHAM GOTTLOB (1749-1817)

Kurze Klassifikation und Beschreibung der verschiedenen Gebirgsarten (Short Classification and Description of the Different Rocks. Dresden; In Der Watherischen Hofbuchhandlung

Classified rocks as formations instead of individual rock types
GRA 38

Classified rocks by their respective ages, worldwide
FRMS 201

1788

BEROLDINGEN, FRANZ COELESTIN VON (1740-1798)

Bemerkungen auf einer Reise durch die pfalzischen und zweybruckschen Quecksilber-Berquerke. Berlin; F. Nicolai

Said quicksilver deposits are due to sublimation from plutonic vapors
HTOD 40

MONNET, ANTOINE GRIMOALD (1734-1817)

Mémoire sur la Formation des Minéraux, in Académie des Sciences, Paris. Mémoires: 3: 337-56.

Said mineral formation involved the conversion of water into earth

MCR 59

1789

LAVOISIER, ANTOINE LAURENT (1743-94)

Observations Générales, sur les Couches Modernes Horizontales, in Académie des Sciences, Paris. Mémoires: 351-71

Said adjacent but different sedimentary environments resulted from changes
CONK 12; D&B 48; GEI 344

MITCHILL, SAMUEL LATHAM (1764-1831)

Geological Remarks on Certain Maritime Parts of the State of New York, in American Museum (Universal Magazine), 5: 123-26

Said East Coast had not all been sea bed but was granite sculpted by erosion in some areas
HAZ 273-4

RAZOMOVSKY, GREGOR DE (1759-1837)

Histoire Naturelle du Jorat et de ses Environs. Lausanne; J. Mourer

Said waters of early earth were very stro[ng] and carried huge boulders long distances
HOOY 279

VANNUCCI, GIUSEPPE

Discorso Istorico-Filosofico sopra il Tremuoto. Cesena; G. Biasini

Said the Rimini earthquake had been cause[d] by several days of lightning which occurred just before the disturbance
AD 412

WERNER, ABRAHAM GOTTLOB (1749-1817)

Herrn D. Fausts Nachricht von dem auf dem Meissner in Hessen ueber Steinkohlen und bituminosem Holze liegendem Basalte, in Bergmaennisches Journal, 1: 261-95

Said basalt was a precipitate
SEDD 390

Versuch einer Erklarung der Entstehung der Vulkanen durch die Entzundung mächtiger Steinfohlenschichten, als ein Beytrag zur Naturgeschichte des Basalts, in Höpfner's Magazin fur die Naturkunde helvetiens, 4: 135

Said volcanoes usually are result of underground coal catching fire and did no[t] exist before the organic-bearing floetz rocks formed
GEI 225, WOOD 34

WILLIAMS, JOHN (1730-95)

Natural History of the Mineral Kingdom. Edinburgh; T. Ruddiman

Said earth was originally a fluid mass which mountain-high tides spread into regular layers, after which heavy materials settled to the bottom, and diluvian waters then reduced the strata to conglomerate finally tilting the remnants thus creating topography
MILL 174; NEILL 82

Said escarpments were formed by regularly dipping strata
EPBG-III 113

Said veins were filled by mineral-rich waters, while coal formed in basins, and volcanics formed where lava accumulated
NEILL 82-3

179?

HUTTON, JAMES (1726-97)

Principles of Agriculture (never published)

Said the genetic features that are best adapted for a breed will carry on in the line and will continue to perfect itself through natural variations which continually occur
BAIL 16-7

1790

HUMBOLDT, FRIEDERICH HEINRICH ALEXANDER VON (1769-1859)

Mineralogische Beobachtungen ueber einige Basalte am Niederheim (Basalts of the Rhine)

Said basalt was aqueous in origin
AD 237

HUTTON, JAMES (1726-1797)

Observations on Granite, in Royal Society of Edinburgh. Transactions, 3: 77-85 (1794)

Said granite had consolidated from molten matter throughout time
AD 242; BAIL 28; GNC 31

LUC, JEAN ANDRE DE (1727-1817)

Letters to Dr. James Hutton, F.R.S., Edinburgh, on His Theory of the Earth, in Monthly Review, New Series, 2: 206-27, 582-601, 3: 573-86 and 5: 56

Said soil was not carried away by rain, streams, etc., nor were mountains eroded to river-carried gravels
CHOR 48-9

Said present continents, recently formed, were in their original form and were not weathering away
PORT 165

1790?

HUNTER, JOHN (1728-93)

Observations and Reflections on Geology. London; Taylor and Francis (1859)

Said a shifting of the polar axis changed the angle at which the sun's rays hit the earth and so changed climates
JONES 237

Said fossil forms changed little through hundreds of years and the surrounding rock was the same age as the fossils it contained
JONES 220, 236, 238-9

Said land had been covered by the sea several times
JONES 235

Said many fossils were of extinct species but there was no progression and many land fossils were buried beneath the ocean now
JONES 237, 240

Said most of the earth´s inanimate matter was generated by an "agency of life" of which every animal contains a portion
JONES 241

Said the major agent of topography was water, over time, both as sculptor and as transporter of land into the ocean
JONES 235, 242-3

Said volcanoes raised mountains and built islands
JONES 242

Said waves carved shoreline and piled up sediment on ocean floors to form submarine mountains
JONES 243

1791

BEDDOES, THOMAS (1760-1808)

Observations on the Affinity between Basaltes and Granite, in Royal Society of London. Philosophical Transactions, 81: 48-70

Said basalt and other crystalline rocks were of igneous origin
PORT 175

Said granites were metamorphic in origin
SARJ 184-6

DOLOMIEU, (DEODAT) GRATET DE (1750-1801)

Mémoire sur les Pierres Composées et sur les Roches, in Observations sur la Physique, sur l´Histoire Naturelle et sur les Arts, 39: 222-39, 374-407 and 40: 41-62, 203-18, 372-403

Said the erratics had been moved over slightly inclined slopes by a series of large marine floods
GIA 158

Classified rocks by origin (precipitate, sedimentary, or volcanic)
HSP 2

Said each simple earth had formed during a different period of the earth´s history, ordered according to their relative solubilities
OLD RHG 235

Said slow sedimentation from the ocean laid down horizontal primitive rocks which a world wide catastrophe then disrupted thus creating mountains from which future inundations of the sea created the present topography
HOOY 281

WERNER, ABRAHAM GOTTLOB (1749-1817)

Neue Theorie von der Antstehung der Gange (New Theory of Mines). Freiberg; Gerlachische Buchdrukkerei

Inserted a transitional unit (limestone, graywacke, and trap rock) between the unfossiliferous primitive rocks and the fossiliferous floetz
ALB-2 121-2

Said veins were not rocks and those whose strike trends were parallel were of the same age, while one which crossed another was younger
GEI 309; HTOD 61; SEDD 388

Said veins were result of cracks in rocks being filled by solutions flowing from waters above as the waters themselves were precipitating the rocks
CHOR 28-9; GEI 309; HTOD 35; TODH 319

1792

HAUY, RENE JUST (1743-1822)

De la Structure, Considérée comme Caractère Distinctif des Minéraux, in Journal d´Histoire Naturelle, 2: 54-65 and in Paris, Séances Ecoles Normales, Appendix: 79-92

Said elementary particles combined into a crystal form unique to a species thereby determining color, shape, etc.
BJG 108

1792-97

SPALLANZANI, LAZZARO (1729-99)

Viaggi Alle Due Sicilie ed in Alcuni Parti dell´ Appennino Pavia (Travels in the Two Sicilies and Some Parts of the Appenines). Pavia; Stamperia di B. Comini

Said there was water, as well as volatiles important in volcanic activity, in magma
HSP 60, 73

1793

FRANKLIN, BENJAMIN (1706-90)

Queries and Conjectures Relating to Magnetism, and the Theory of the Earth, in American Philosophical Society. Transactions, 3: 10-13

Said earth had a gaseous core
BRUSH 708

Said tropical fossils found in the north were explainable by shifts in the iron-rich fluid interior which caused the poles to move
HAZ 189-90

HAUY, RENE JUST (1743-1822)

Exposition de la Théorie sur la Structure des Cristaux, in Annales de Chemie, 17: 225-319

Said crystal faces were built by the successive addition of lamellae
BJG 94; SMAJ 12

LUC, JEAN ANDRE DE (1727-1817)

Geological Letters Addressed to Professor Blumenbach, in British Critic, 2+: 231+

Said as precipitation lowered sealevel, land with animals and man appeared, but the land collapsed and a universal deluge occurred, after which the present land rose from the old sea bed and remained, unchanging and permanent
PORT 166, 200-1

Said early earth was a chaotic fluid from which precipitated the unfossiliferous primary strata, and then the secondary strata in which living marine creatures were imbedded
PORT 166, 200-1

MARTIN, WILLIAM (1767-1810)

Figures and Descriptions of Petrifications Collected in Derbyshire. Wigan; W. Lyon

Named fossils in the Linnean fashion but used a trinomial system
PBG 194

WITT, BENJAMIN DE (1744-1819)

(Letter), in Transactions of the Philadelphia Academy, Volume 2

Said erratics had been erupted or ejected from the earth by earthquakes
MERR 211

1794

CHLADNI, ERNST FLORENS FRIEDRICH (1756-1827)

Ueber den Ursprung der von Pallas gefundenen und anderer ihr ahnlicher Eisenmassen, und ueber einige damit in Verbindung stehende Naturescheinungen. Riga; N.P.

Said earth and meteorites consisted of an analogous chemical composition
TIK 327

HUNTER, JOHN (1728-93)

Account of Some Remarkable Caves in the Principality of Bayreuth and of the Fossil Bones Found Therein, in Royal Society of London. Philosophical Transactions: 402-17

Said land and sea had changed places in the past
PORT 160

SULLIVAN, RICHARD JOSEPH (1752-1806)

View of Nature. London; T. Becket

Said change in nature was by very slow gradations
PORT 233

Said strata of land and sea formed at different times and over a very long period
PORT 158-9

1794-95

DESMAREST, NICHOLAS (1725-1815)

Encyclopedie Méthodique, ou par Ordre de Matières: Par une Société de Gens de Lettres, de Savans et d'Artistes: Géographie Physique, Volume 1-5 Paris; Chez H. Agasse

Said strata consisted of 1) ancient granite solidified out of magma melted by subterranean burning coal and possibly preceded by cycles of creation and destruction, 2) secondary, and 3) tertiary deposits
TND 350

1794-96

DARWIN, ERASMUS (1731-1802)

Zoonomia. Dublin; P. Byrne and W. Jones

Said evolution was the result of new characteristics which survived and were passed on
D&B 63

Said evolutionary change was the culmination of many forces
D&B 63

Said species adapted to their environment primarily by their food gathering techniq[ues]
D&B 63

1795

HUTTON, JAMES (1726-97)

Theory of the Earth, with Proof and Illustrations. Edinburgh; Cadell and Davies

Said the oily matter which seeped up into the springs was distilled from coal, or the material that formed coal, and rose with water, through fissures
OWEN 17

Said cliffs were eaten away when rivers´ load plus debris from cliff was removed by waves but, when they exceeded waves´ capacity a ´shelving shore´ would form
BAIL 114

Said effect of heat on rocks could be modified by pressure
GEI 301, 310-11

Said rivers wore off rough edges of rocks moved downstream but the coastal ocean waves are the most effective rounders
BAIL 105

Said strata, including coal, had been sediments which drifted together on the ocean floor and then consolidated
BAIL 92; GNC 20

Said topography of emergent land will be the result of 1) original stratification, 2) deep-seated forces, and 3) weathering
BAIL 98

Said that mountain uplift was caused by heat beneath the crust of the earth
GER 27

Said alpine glaciation had been widespread at one time when the Alps were much higher
D&B 425; GIA 159

Said composition of stratified rock was altered by heat and pressure, during consolidation and while buried in the earth
GEI 310-1; GNC 20

Said glaciers had carried detritus, including granite blocks, great distances down the mountain slopes
BAIL 111-2; GEI 314; GIA 159

Said horizontal beds of strata form table mountains
CHOR 53

Said river channels become larger as they merge and decrease in number, and may form a system of rivers and valleys complementing each other
CHOR 53, 63

Said streams flow around resistant hills but eventually wear through them and leave gaps when they erode bedrock
CHOR 93

Said structure determined the form of a mountain
EPBD-III 123

Said that some rocks partially melted and then cooled into a mixture of different rocks which formed bedrock
HCM 51, 54

Said that the earth´s original rocks had been altered by heat and pressure
HDMF 149

Said 1) high mountains have young rivers with heavy loads, and 2) as mountains weather, load decreases and valley erosion increases downstream, until 3) flat plains result
CHOR 52

THOMSON, WILLIAM (GUGLIELMO) (1761-1806)

Abbozzo di una Sciagrafia Volcanica, in Giornale de´ Letterati di Napoli, 41: 59-67

Classified lavas by their external characteristics as 1) vitreous, 2) pitch-stone, 3) crystalline, and 4) earth
WATER 126

1796

CUVIER, GEORGES (1769-1832)

Work	Contribution
Discours sur les Révolutions de la Surface du Globe, in Recherches sur les Ossemens Fossiles de Quadrupeds. Paris; Chez Deterville (1812)	Said certain fossils were bones of extinct species HMPD 338

JAMESON, ROBERT (1774-1854)

Work	Contribution
Is the Volcanic Opinion of the Formation of the Basaltes Founded on Truth? in Dissertations of the Royal Medical Society, 35 (20): 218-24 (1795-96)	Said basalts sometimes contained fossils and were sedimentary rocks RJA 87

LAPLACE, PIERRE SIMION (1749-1847)

Work	Contribution
Exposition du Système du Monde. Paris; Circle-Social	Said early gaseous universe condensed and contracted, particles formed on edges with sun toward center, and the masses of particles collected, eventually forming planets MATH 143; MHG 199; P&S 8

SAUSSURE, HORACE BENEDICT DE (1740-99)

Work	Contribution
Voyages dans les Alpes. Neuchatel; S. Fauche	Said geodes and calcareous concretions formed in successive layers around a nucleus or in a cavity AD 130
	Said limestone mountains formed horizontally by mechanical deposition and later were folded while in a softened, plastic state AD 390
	Said pebbles and gravel were debris from rocks broken down, transported and rounded by running water AD 129-30

SMITH, WILLIAM (1769-1839)

Work	Contribution
Memoirs of William Smith, edited by John Phillips. London; J. Murray (1844)	Said each stratum had its unique fossils CONK 13; D&B 22; FENT 83; PBG 179

1797

BERTRAND, PHILIPPE M. (1730-1811)

Nouveaux Principes de Géologie. Paris; Chez l´Auteur

Said frozen globe was struck by three comets; one melted it and water life appeared, the second displaced the water and land appeared and the third comet caused volcanoes and earthquakes and thus, topography
MILL 178-80

BUCH, CHRISTIAN LEOPOLD VON (1774-1853)

Versuch einer mineralogischen Beschreibung von Landeck, in Gesammelte Schriften. Berlin; G. Reimer (1867-85)

Said basalt was aqueous in origin and sometimes contained organic material
GEI 246

FOND, BARTHELEMY FAUJAS DE ST. (1742-1819)

Voyage en Angleterre en Ecosse et aux Iles Hébrides. Paris; J. H. Jansen

Said some basalts were volcanic in origin
WOOD 22

HOWARD, PHILIP (??-1810)

Scriptural History of the Earth and Mankind. London; R. Faulder

Said topography of earth was unchanging
PORT 165

LAMARCK, JEAN BAPTISTE PIERRE ANTOINE DE MONET DE (1744-1829)

Tableau Encyclopedie et Méthodique des Trois Regnes de la Nature. Paris; Panckducke (1791-1823)

Said crystals formed from decayed organic matter
FRMS 200

METHERIE, JEAN CLAUDE DE LA (1743-1817)

Théorie de la Terre. Paris; Maraden (1795)

Said earth´s crust solidified under water, complete with present mountains and valleys, after which the waters drained into the earth
MILL 176-7

1798

HALL, JAMES (1761-1832)

Experiments on Whinstone and Lava, in Nicholson Journal (Journal of Natural History, Chemistry and the Arts), 4: 8-18, 56-65 and Royal Society of Edinburgh, Transactions, 5: 43

Said lava rose and filled dykes, heating the country rock which was then altered, and fast cooling lava near the rock bcame vitreous while its slower cooling exterior became crystalline
GEI 321-2

Demonstrated experimentally that slow cooling produced stony, crystalline textures and fast cooling produced a glassy texture
WOOD 40-1

HATCHETT, CHARLES (1765-1847)

Analysis of the Earthy Substance from New South Wales, Called Sydneia or Terra Australis, in Nicholson Journal (Journal of Natural Philosophy, Chemistry and the Arts), 2: 72-80 (1799)

Said bitumin was produced by the decomposition of plant and animal fats, oils and resins
OWEN 17

LATROBE, BENJAMIN HENRY (1764-1820)

His Journal, 30 April 1798, in Microfiche Edition. Clifton, NJ; James T. White and Co. (1976)

Said rocks weathered to form soil
BHL 110

Suggested the westward tilt of North America was caused when the moon was ejected from the Pacific area thus creating the ocean basin and triggering uplift of some marine strata (he then disavowed himself of this idea)
BHL 109

Said stream erosion cut small valleys and only slightly shaped the land
BHL 109

THOMSON, WILLIAM (GUGLIELMO) (1761-1806)

Sur l´Origine de l´Oxigène Nécessaire pour Entretenir le Feu Souterrain de Vésuve, in Giornale de´ Letterati di Napoli, 106: 3-46

Said volcanoes decomposed limestone and its constituent carbonic acid, thus liberating oxygen to fuel the fire
WATER 126

TOWNSON, ROBERT (1763-97)

Philosophy of Mineralogy. London; Sold by J. White

Said earth had undergone multiple evolutions of formation and change
PORT 158

1799

BERTRAND, LOUIS (1731-1812)

Renouvellemens Périodiques des Continents Terrestres. Paris; C. Pougens

Said comet struck earth, displacing the rotating magnetic nucleus inside and causing seas to cover old continents and to uncover present ones
MILL 181

DALTON, JOHN (1766-1844)

Experiments and Observations to Determine Whether the Quantity of Rains and Dew Is Equal to the Quantity of Water Carried Off by the Rivers and Raised by Evaporation, with an Enquiry into the Origin of Springs. Manchester; R. and W. Dean and Co.

Said rain and dew provide as much water as is carried away by rivers and evaporation
AD 449

KIRWAN, RICHARD (1733-1812)

Geological Essays. London; D. Bremner

Said coal was remains of primitive carbon mountains weathered away by rain, etc.
PORT 160

Said early earth solidified elements in layers according to their elective attraction, and these chemical reactions released heat which fractured the strata, freeing oxygen to form an atmosphere, permitting life to arise on newly uplifted land
MILL 182-3

Said warm climate fossils had been carried to Ireland by the flood
CGT 3

SMITH, THOMAS P. (1775?-1830)

Account of Crystallized Basalts Found in Pennsylvania, in American Philosophical Society. Transactions, 4: 445-6.

Said crystallized basalts were aqueous in origin
FGA 1828

SMITH, WILLIAM (1769-1839)

Tabular View of the Order of Strata in the Vicinity of Bath with Their Respective Organic Remains. Never published (Manuscript in the British Museum)

Said that each English stratum had its own fossils which were unlike any others and could be used to identify that strata
GEI 389; WOOD 42

1800

BRONGNIART, ALEXANDRE (1770-1847)

<u>Chalk of Rouen and of the Roches des Fiz in the Arve Valley</u>. N.P.; N.P.

Used fossils to correlate strata over great distances
MARG 101

1801

BRUNNER, JOSEPH (1764-1807)

<u>Neue Hypothese von Entstehung der Gänge</u>. Leipzig; Andra

Said primitive molten rock was composed of three primitive constituents
AD 321

HAUY, RENE JUST (1743-1822)

<u>Traité de Minéralogie</u>. Paris; Chez Louis

Classified minerals chemically into four classes, including orders, genus and spec
SMAJ 15

Devised a shorthand notation for faces of crystals composed of capital letters (faces), vowels (angles), and consonants (edges)
BJG 94-100

Said combinations of six basic geometrica forms, determined by their chemical components, accounted for all crystal shapes
FRMS 195

Said rocks were aggregates of 1) volcanic minerals, or 2) nonvolcanic minerals, and were classifiable by age
CIR 42; HTOD 143-4

Said the angles formed by the plane faces of like crystals were in simple integral ratios, a definite mathematical relation-ship
BJG 78; SCH 15

Said identification of minerals depended 1) on crystallography, and then 2) on chemistry
KUNZ 71; WMC 67

HUMBOLDT, FRIEDERICH HEINRICH ALEXANDER VON (1769-1866)

Sketch of a Geological Delineation of South America, in Philosophical Magazine, 17: 347-57 (1803)

Said Atlantic Ocean was a valley dug out by a torrential current
CDBN 2

LAMARCK, JEAN BAPTISTE PIERRE ANTOINE DE MONET DE (1744-1829)

Système des Animaux sans Vertèbres. Paris; Chez Deterville

Said earth was very old
AD 268; GEI 358; WOOD 52-53

Said fossil forms varied with changes in their natural environments
WOOD 52-3

Said fossil shells were mostly marine organisms, from differing depths, which had accumulated naturally in sediments which were later exposed
WOOD 52-53

Said fossils and topography were both the results of natural, presently occurring, events
CHA 335

Said new species descended from previous ones
HMPD 341

Said that living organisms changed through time, and given long enough, all variations were possible; however, species were a man-made distinction and only gradation, never extinction, was the rule
GEI 358; RUD 119

1802

BUCH, CHRISTIAN LEOPOLD VON (1774-1853)

Geognostische Beobachtungen auf Reisen durch Deutschland und Italien. Berlin; Haude (1802-09)

Said mountains were underlain by vast caverns with lava walls
AD 384

Said basalt was aqueous in origin
FENT 74

Observations sur les Volcans d´Auvergne, in Journal des Mines, 13: 249-56 (1802-03)

Said basalts of Auvergne were volcanic in origin
GEI 247

LAMARCK, JEAN BAPTISTE PIERRE ANTOINE DE MONET DE (1744-1829)

Hydrogéologie. Paris; L'Auteur	Said migrating oceans caused the poles and earth's center of gravity to shift GEI 358
	Said the moon and sun cause tides whch trap detritus near shore, preventing it from entering the ocean depths and causing overflow and thus building new land GEI 352-5
	Said nonvolcanic mountains were formed by water cutting away plains AD 268
	Said that the inclined strata of mountains had been a sloping seashore or was the result of local subsidence CHOR 81-2
	Said the sea had progressively covered different lands for vast periods AD 268
	Said when organisms die, they decay and their constituents become parts of new minerals GEI 360
	Said that plains are cut by valleys and highlands are worn down by alternating drought and flood, heat and cold, and running water GEI 352-5

PLAYFAIR, JOHN (1748-1819)

Illustrations of the Huttonian Theory of the Earth. Edinburgh; W. Creech	Said heat combined with upward pressure caused folded strata GKES 148-9
	Said a river's load would wear away cataracts (i.e., nickpoints) CHOR 64
	Said erratics had been moved by glaciers and rivers FENT 68; GEI L&P 198
	Said folded strata was the result of gravity and resistance intruding lateral and oblique movement to the force which uplifts strata BEL 31

Said incised meanders could not have been cut by a flood
CHOR 64

Said lakes were formed 1) when soluble materials were removed or 2) as a result of uplift or subsidence
GEI L&P 199

Said plains bordering primitive mountains were composed of river-carried stone from the mountains, varying in size depending on the distance carried, being broken down and rounded the further away the source was
CHOR 63

Said rivers and tributaries were individually balanced with respect to grade, velocity, volume, and payload to remove irregularities and form a mutually integrated system
CHOR 62

Said rivers cut their own valleys with varying gradients, depending on the varying resistances of the rocks involved
CHOR 63-4

Said that when rivers deposited their load in the sea, the heaviest materials settled and lighter materials were carried farther downstream
CHOR 63

Said the alpine glaciers had formed one great ice field and such a glacier could move huge rocks, such as the erratics
CGT 17; DGGS 74; FOREL 108-9; GEI 442; GIA 159

803

UBUISSON, JEAN FRANCOIS DE VOISIONS (1769-1819)

Sur les Basaltes de la Saxe, in Annales de Chimie, 46: 170-90, 225-50

Said basalts of Saxony were aqueous in origin
GEI 242

FOND, BARTHELEMY FAUJAS DE ST. (1742-1819)

Essai de Géologie ou Mémoires pour Servir à l'Histoire Naturelle du Globe. Paris; Chez G. Durour et Co. (1809)

Said fossils could eventually be used to date strata
CBS 218

Said that metals and their ores grow much as plants do in the earth
AD 296

LUC, JEAN ANDRE DE (1727-1817)

Abrégé de Principes et de Faits Concernans la Cosmologie et la Géologie. Brunswic; Maison des Orphelins

Said frost would erode bare surfaces but could not work through vegetation
BAIL 109-10

VOLNEY, CONSTANTIN FRANCOIS CHASSEBOEUF (1757-1820)

Tableau du Climat et du Sol des Etats-Unis d'Amérique (View of the Soil and Climate of the United States of America). Paris; Courcier et Dentu/ Philadelphia; J. Conrad & Co. (1804)

Said earthquakes occurred often in the eastern United States because of a layer of schist that ran northward to Lake Ontario, the crater of a large volcano
FENT 129-30

1804

AUBUISSON, JEAN FRANCOIS DE VOISIONS (1769-1819)

Sur les Volcans et les Basaltes de l'Auvergne, in Société Philomatique de Paris Bulletin, 3: 182-5

Said basalts of Auvergne and Saxony were of igneous origin
FENT 73; GEI 244-5

PARKINSON, JAMES (1775-1824)

Organic Remains of a Former World, Volumes 1-3. London; C. Whittinghan

Said fossils were remains of species living before the deluge destroyed everything
HMPD 342

SCHLOTHEIM, ERNST FRIEDRICH VON (1764-1832)

Beschreibung merkwurdiger Krauter-Abdrucke und Pflanzen-Versteinerungen. Ein Beitrag zur Flora der Vorwelt. Gotha; Becker

Said carboniferous age fossils were the remains of extinct flora
RUD 146

THOMSON, WILLIAM (GUGLIELMO) (1761-1806)

Essai sur le Fer Malléable Trouvé en Sibérie par le Prof. Pallas (On Malleable Iron), in Bibliothèque Brittanique, 27 (2-3): 135-54, 209-29

Described and sketched etching of meteoritic iron, the triangular pattern, the constancy of angles, and the differing solubility of the three components
WATER 129-30

1805

DAVY, HUMPHRY (1778-1829)

Lecture on the Theories of the Origins of the Rocks, in Lectures Presented at the Royal Institution in 1805. Manuscript kept at the Royal Institution in London

Said some basalts had formed after melting but some had never been molten
SIEG 223

Said volcanoes resulted from chemical reactions within the earth
NCDAFG 229

OKEN, LORENZ (1779-1851)

Abriss der Naturphilosophie (Natural Philosophy). Gottingen; VandenHoek und Ruprecht

Said fissures resulted from loss of water of crystallization and metals formed in outer crust in the presence of air, earthy water and darkness
TODH 319

HALL, JAMES (1761-1832)

Account of a Series of Experiments Showing the Effects of Compression in Modifying the Action of Heat, in Royal Society of Edinburgh Transactions, 6: 71-183 (1812)

Said crystallization was result of fusion at at high temperatures
BJG 28-9; GEI 233-4

1806

BARTON, BENJAMIN SMITH (1766-1815)

Facts, Observations, and Conjectures, Relative to the Elephantine Bones (of Different Species), That Are Found in Various Parts of North America, in Philadelphia Medical and Physical Journal, First Supplement 22-35

Said the mammoth was herbivorous and extinct
HAZ 70

1807

BRONGNIART, ALEXANDRE (1770-1847)

Traité Elémentaire de Minéralogie. Paris; Deterville

Classified minerals into five classes including orders, genus and species
SMAJ 15

LATROBE, BENJAMIN HENRY (1764-1820)

Letter, in Papers Relative to the Completed Bridge across the Potomac. Washington, D.C.; Duane & Son

Said clearing of land near rivers was speeding up erosion and sedimentation
BHL 109-10

YOUNG, THOMAS (1773-1829)

Lecture on Natural History, in Course of Lectures on Natural Philosophy and the Mechanical Arts. London; J. Johnson

Said earthquakes were waves of vibration
DAV 68

1807-15

FREIESLEBEN, JOHANN CARL (1774-1846)

Geognostischer Beytrag zur Kenntnis des Kupferschiefergebirges mit besonderer Hinsicht auf einen Theil der Graffschaft Mannsfeld und Thüringiens, in Geognostische Arbeiten. Freyberg; Craz und Gerlach

Separated formations based on their fossils and lithology
RUD 129

1808

CUVIER, GEORGES AND BRONGNIART, ALEXANDRE (1769-1832 / 1770-1847)

Essai sur la Géographie Minéralogique des Environs de Paris (Essay on the Mineral Geography of the Paris Region), in Annales du Museum d'Histoire Naturelle, Paris, 11: 293-326 and in Journal of Mines, 23: 426-27, 434-35

Said the life history of an area was revealed by its minerals, stratigraphic sequences, and fossil assemblages, each of which were unique to given strata
CBS 219; RUD 129; WOOD 55

Differentiated between multi-bed stages, or deposits
CONK 37

Said strata were uniform and were distinguished by a succession of fossil types, the most recent of which were similar to living creatures
CONK 37

Described a discontinuity and said it represented a long period of time
CONK 12, 37

Said changes occurred differently in the ancient seas
CONK 37

Said alternation between marine and fresh water strata suggested a cyclic rising and falling of the land
OLD RHG 236; RUD 130

DALTON, JOHN (1766-1844)

A New System of Chemical Philosophy, Volume 1-2. Manchester; S. Russell for R. Binkerstaff

Said crystal shapes were determined by the stacked arrangement of their spherical units of matter
SMAJ 20

Said the shape created by crystallization revealed the natural arrangement of the particles composing the compound substance
BJG 120

JAMESON, ROBERT (1774-1854)

Elements of Geognosy, in System of Mineralogy. Edinburgh; Longman, Hurst, Rees and Orme (1804-08)

Said obsidian and pumice were aqueous precipitates
GKES 138

Said inclined strata had been laid down on an incline
GKES 149

1809

CARR, JOHN

On the Natural Causes which Operate in the Formation of Valleys, in Philosophical Magazine, 33: 452-9 and 34: 190-200

Said mountains were left when less durable rock was washed away
CHOR 80-1

Said valleys were carved by their rivers
CHOR 80-1

CLEAVELAND, PARKER (1780-1858)

Account of Fossil Shells, with the Author's Reasons for Attending to the Same, in American Academy of Arts and Sciences Memoirs, Volume 3, Part 1: 155-15

Said fossils could be used to correlate strata in different areas
HAZ 70-71

Said the great number of fossils found worldwide were caused by changes wrought in the earth´s interior
FGA 1830

EBEL, JOHANN GOTTFRIED (1764-1830)

Ueber den Bau der Erde in den Alpen-Gebirge zwischen 12 Langren- und 2-4 Breitengraden nebst einigen Betrachtungen ueber die Gebirge und den Bau der Erde überhaupt mit geognostischen Karten. Zurich; Orell Fussli

Said erratics had been transported over gently inclined surfaces by a series of inundations of the ocean
GIA 158

FAREY, JOHN (1776-1826)

Geological Observations on the Excavation of Valleys and Local Denudations of the Strata of the Earth in Particular Districts, in Philosophical Magazine, 33: 442-8

Said faults were the result of the earth´ surface rising toward a dense body moving through space near the earth
CHOR 79

Said outliers appeared when a great denud tion occurred
CHOR 79

GODON, SILVAIN (1774?-1840)

Mineralogical Observations Made in the Environs of Boston, in Memoirs of the American Academy of Arts and Sciences, 3: 127-154

Said minerals were 1) simple (acid, earth combustible, and metallic) or 2) aggregat (primordial soil and alluvial) and rocks should be named by adding ´oid´ to the most characteristic mineral present
FOHYAG 37

MALUS, ETIENNE LOUIS (1775-1812)

Sur une Propriété de la Lumière Réfléchie par les Corps Diaphanes, in Société d´Arcueil Mémoire de Physique et de Chimie, 2: 143-58

Described the polarization of light
SMAJ 18

MARTIN, WILLIAM (1767-1810)

Outlines of an Attempt to Establish a Knowledge of Extraneous Fossils. Macclesfield; J. Wilson

Said fossils were key to dating strata and to explaining the formative processes of minerals
PORT 167

1810

BUCH, CHRISTIAN LEOPOLD VON (1774-1853)

Etwas ueber Locale und allgemeine Gebirgsformationen, in Gesellschaft naturforschender Freunde Magazin, 4: 69-74	Said erratics were flood carried and received striations when scraped along the bottom CHOR 198

DAY, JEREMIAH (1773-1867)

View of the Theories Which Have Been Proposed, to Explain the Origin of Meteoric Stones, in Connecticut Academy of Arts and Sciences Memoir 1, Part 1: 163-74	Said when eccentric orbits carried comets too close to earth, they were drawn to earth HAZ 226

LUC, JEAN ANDRE DE (1727-1817)

Geological Travels in Some Parts of France, Switzerland and Germany. London; F. C. and J. Rivington (1813)	Said some erratics were ejected almost in place by gaseous eruptions of sinking strata forming valleys GIA 158

1811

AIKEN, ARTHUR (1773-1854)

Observations on the Wrekin, and on the Great Coal Field of Shropshire, in Geological Society of London Transactions, 1: 191-212	Said petroleum could not originate from coal OWEN 31-2

BREISLAK, SCIPIONE (SCIPION) (1748-1826)

Introduzione alla Geologia. Milano; Stamperia Reale	Said that metal elements were precipitated according to their specific gravity, from the early molten earth and migrated into fissures to form veins of physically and chemically similar minerals TODH 319
	Said igneous rocks resulted from combination of heat and water HTOD 41

NUGENT, NICHOLAS (1781-1843)

Account of a Pitch Lake of the Island of Trinidad, in Geological Society of London Transactions, 1: 63-76	Said petroleum and the byproduct pitch were the result of volcanic action affecting buried organic material OWEN 30

1811

PARKINSON, JAMES (1755-1824)

Observations on Some of the Strata in the Neighborhood of London, in Geological Society of London Transactions, 1: 324-6

Said fossils might be washed out of earlie deposits or share a common boundary thus causing an incorrect date to be assigned to the strata
CONK 38

PINKERTON, JOHN (1758-1826)

Petrology: A Treatise on Rocks. London; S. Hamilton and Co.

Classed minerals into six ´substantial´ and six ´circumstantial´ classes: 1) siderous, 2) siliceous, 3) argillaceous 4) magnesian, 5) calcareous, 6) carbonaceous, 7) composite, 8) diamictonic, 9) anomalus, 10) transilient, 11) decomposed, and 12) volcanic
CROSS 343-4

1812

BIOT, JEAN BAPTISTE (1774-1862)

Mémoire sur un Nouveau Genre d´Oscillation que les Molécules de la Lumière Eprouvent en Traversant Certains Cristaux, in Institut des Sciences, Lettres et Arts. Mémoires Présentés Par Divers Savants, 1: 1-371

Described optical biaxiality
SMAJ 18

BREWSTER, DAVID (1781-1868)

On the Affections of Light Transmitted through Crystallized Bodies, in Royal Society of London. Philosophical Transactions, 1: 187-218

Described optical biaxiality
SMAJ 18

BUFFON, GEORGES LOUIS LECLERC (1707-1788)

Historie Naturelle (Second Edition), in Volume 1: 128-32. Paris; L´Imprimerie Royale

Said there had been a flood but it was not violent
WDC 264

CUVIER, GEORGES (1769-1832)

Discours Préliminaire sur la Théorie de la Terre, Volume 1, in Recherches sur les Ossements Fossiles de Quadrupedes. Paris; Chez Deterville

Said earth had had long, peaceful history punctuated by sudden changes in physical conditions effected by no known process
RGIF 337; RUD 132

Said there was a progression, not necessarily evolutionary, observable in the fossil record
PORT 167; RUD 143

Discours sur les Révolutions de la Surface du Globe, in Recherches sur les Ossements Fossiles de Quadrupedes. Paris; Chez Deterville

Said species died and strata were destroyed by catastrophic floods and overflowing of the oceans, coupled with upheavals of land, followed by quiet periods
AD 232, 266; CLOUD 37; GEI 373-4; RGIF 338

Said fossils revealed a succession of different series of species
HMPD 339

Said when species were destroyed, new ones, more advanced, appeared in finished form
HUBB 232; GEI 374-5

Said an entire organism could be described from any individual bone
HMPD 339

Recherches sur les Ossemens Fossiles de Quadrupedes. Paris; Chez Deterville

Said fossils suggested the age of the strata
AD 265; D&B 22

Sur un Nouveau Rapprochement à Etablir entre les Classes qui Composent le Règne Animal, in Annales du Museum d'Histoire Naturelle, 19: 73-84

Said the animal kingdom consisted of 1) vertebrates (fish to mammal) and 2) molluscs (Acephala to cephalopods) with four classes each
RUD 142-3

HALL, JAMES (1761-1832)

On the Revolutions of the Earth's Surface, in Royal Society of Edinburgh Transactions, 7: 139-212+ (1813)

Said striations in rocks were caused by large currents of rushing water
CGT 10

Said violent uplift of an island might produce a wave capable of washing boulders off peaks of mountains and depositing the erratics in distant lands
CGT 13

WOLLASTON, WILLIAM HYDE (1766-1828)

On the Primitive Crystals of Carbonate of Lime, Bitter Spar, and Iron Spar, in Royal Society of London. Philosophical Transactions 159-62

Said the variability of composition of these crystals suggested that they were mixtures
BJG 119

1813

BAKEWELL, ROBERT (1768-1843)

Introduction to Geology. London; J. Harding

Said cliffs and heights were continuously worn down until they became covered with soil and vegetation which then protected them from further decay
CHOR 78

BRONGNIART, ADOLPHE THEOPHILE (1801-1876)

Essai d´une Classification Minéralogique des Roches Mélangées, in Journal des Mines, 34 (199): 5-48

Classified rocks as 1) cristallisées isomères, 2) crystallisées anisomères, and 3) agregées
HTOD 146-9

COOPER, THOMAS (1759-1839)

Geology, in Emporium of Arts and Sciences, New Series, 3: 412-25

Said earthquakes were caused by expanding gases inside the caverns of the earth
HAZ 190

Said the earth had a solid core of magnetic material, probably iron or nickel
HAZ 190

Said the composition of meteorites might be that of the earth´s interior
HAZ 190

D´HALLOY, JEAN BAPTISTE JULIEN D´OMALIUS (1783-1875)

Observations sur un Essai de Carte Géologique de la France, in Annales des Mines, 7: 353-76 (1822)

Grouped rocks by period of origin as indicated by their fossils saying chemical composition was more variable
CONK 40

DAUXION-LAVAYSEE, JEAN FRANCOIS (1775-1826)

Voyage aux Iles de la Trinité, Tobago et Margarita, et dans Diverses Parties de Venezuela et Amerique du Sud. Paris; F. Schoell

Said igneous activity caused eruptions of mud, sulphurous hydrogen, oil, and asphalt
HIGG 102

SCHLOTHEIM, ERNST FRIEDRICH VON (1764-1832)

Taschenbuch für die gesammte Mineralogie, 3: 2-134
Edited by Karl V. Von Leonhard
Heidelberg; J. C. B. Mohr

Said fossils dated strata relatively and possibly by date
CONK 39; MATH 174

1813

THOMSON, THOMAS (1773-1852)

Some Observations in Answer to Mr. Chenevix´s Attack upon Werner´s Mineralogical Method, in Annals of Philosophy, 1: 241-58

Said it was possible for two minerals to have the same composition but different molecular structure
BJG 112

WOLLASTON, WILLIAM HYDE (1766-1828)

On the Elementary Particles of Certain Crystals, in Royal Society of London. Philosophical Transactions: 51-63

Said crystal shapes were determined by the stacked arrangement of their spherical units of matter
SMAJ 20

1814

AUBUISSON, JEAN FRANCOIS DE VOISIONS (1769-1819)

An Account of the Basalts of Saxony. Edinburgh; A. Constable

Said strata laid down on the sea floor layered the uneven surface, and as peaks emerged as mountains and deposition continued on their still water-covered slopes, eventually it would appear that the higher deposits had thrust up through the lower ones
UHS 388-9

Said that land emerged from the ocean and eroded some, then would again be covered by water where deposition would create new strata at an angle to the old, uneven surface
UHS 389

BERZELIUS, JONS JAKOB (1779-1848)

Versuch durch Anwendung der Electrischen-Chemischen Theorie und der chemischen Proportion-Lehre ein rein wissenschaftliches System der Mineralogie zu Begrunden (An Attempt to Establish a Pure and Scientific System of Mineralogy by the Application of the Electro-Chemical Theory and the Chemical Proportions), in Journal für Chemie und Physik (Schweigger´s Journal), 11: 193-233; 12: 17-62 and 15: 277-300. London; R. Baldwin

Classified minerals solely by chemistry, by the more electronegative element present
WMC 70

BROCCHI, GIOVANNI BATTISTA (GIAMBATTISTA) (1772-1826)

Conchiologia Fossile Subapennina. Milan; Stamperia Reale

Said differences in fossils were due to gradual extinctions following normal patterns of growth and decay during evolution
CONK 39

DAVY, HUMPHRY (1778-1829)

In Memoirs of the Life of Sir Humphry Davy, Volumes 1-2, edited by John Davy. London; Longman, Rees, Orme, Brown, Green and Longman (1790-1868)

Said basalt was a product of lava and was prismatic in the oldest cores
SIEG 223

1815

BAKEWELL, ROBERT (1768-1843)

Introduction to Geology (Second Edition). London; J. Harding

Said rocks encased in ice become lighter and were easily transported from their shoreline location to distant locations
CGT 13

BUCH, CHRISTIAN LEOPOLD VON (1774-1853)

Ueber die Ursachen der Verbreitung grosser Alpengescheibe, in Akademie der Wissenschaften, Berlin. Abhandlungen: 161-86 (1804-11)

Said erratics were carried at different heights within several different marine currents and so were deposited at various elevations
GIA 158-9

CORDIER, PIERRE LOUIS ANTOINE (LOUIS) (1777-1861)

Mémoire sur les Substances Minérales Dites "en Masse" followed by Distribution Methodique des Substances Volcaniques Dites "en Masse", in Journal de Physique, 82: 261-8 (1816)

Said immersion method of microscopic study showed fine grained volcanic rocks were composed of minute quantities of several minerals
CROSS 34-45; GRA 38

KIDD, JOHN (1775-1851)

Geological Essay on the Imperfect Evidence in Support of a Abstract of a Dissertation ... Concerning the System of the Earth, Its Duration and Stability. London; Oxford

Said rivers could cut transverse streams but not alpine valleys
CHOR 81

METHERIE, JEAN CLAUDE DE LA (1743-1817)

Suite à Mes Vues sur l'Action Galvanique comme Cause Principale des Commotions Souterraines et des Volcans, in Journal de Physique, de Chimie, d'Histoire Naturelle et de Arts, 81: 276-87

Said earth solidified from the ocean, not in layers, but crystal on crystal, forming mountains, etc.
AD 388-9

Addition à Mon Mémoire sur les Causes des Commotions Souterraines par l'Action Galvanique, in Journal de Physique, de Chimie, d'Histoire Naturelle et de Arts, 81: 276-87

Said earthquakes were caused by electricity
AD 412

MITCHILL, SAMUEL LATHAM (1764-1831)

A Detailed Narrative of the Earthquake Which Occurred on the Sixteenth Day of December, 1811, in Literary and Philosophical Society of New York Transactions, 1: 281-340

Said several earthquakes and the severe snowstorm of 1810 were caused by a passing comet
FGA 1829

PERRAUDIN, JEAN PIERRE (1767-1858)

Observations Faites par un Paysan de Lourtier. Appended to unpublished manuscript Manuscrit Gillieron de la Bibliothèque Cantonale Vaudoise, Volume 30, by Jean-Siméon-Henri Gillieron

Said striae, polished rocks and boulders high on valley walls were result of ancient, extensive, glaciers
CHOR 193; FOREL 105; GIA 159

SEVERGIN, VASILII MIKHAILOVICH (1765-1826)

Comments on the Probable Age and Origin of Various Mountain Ranges in Russia, in Umozritelnye Issledovania Imperatorsroi St. Peterburgskoi Akademenii Nauk, Volume 4

Said the erratics had been transported by ice
TDG 378

WEISS, CHRISTIAN SAMUEL (1780-1856)

Uebersichliche Darstellung verschiedenen naturlichen Abteilungen der Kristallisations-systeme, in Akademie der Wissenschaften, Berlin. Abhandlungen: 289-344 (1814-15)

Classified crystals as having 1) three or 2) four mutually perpendicular axes
SMAJ 19

1816

CLEAVELAND, PARKER (1780-1858)

Elementary Treatise on Mineralogy and Geology. Boston; Cummings and Hilliard

Classified minerals by chemical composition alone: 1) non-metallic substances, 2) earthy compounds, 3) combustibles, and 4) ores
FGA 1831; HAZ 149; FOHYAG 42; SMAJ 84-5

Classified rocks by mode of creation: 1) volcanic, 2) aqueous, 3) alluvial, 4) primitive, and 5) transitional
HAZ 224; SMAJ 84-5

Said basalt was a secondary rock of aqueous origin
SMAJ 86

DANIELL, JOHN FREDERIC (1790-1845)

On Some Phenomena Attending the Process of Solution and Their Application to the Laws of Crystallization, in Quarterly Journal of Science, 1: 24-49

Said crystal shapes were determined by the stacked arrangement of their spherical units of matter
SMAJ 20

SMITH, WILLIAM (1769-1839)

Strata Identified by Organised Fossils, in Geological Lecture-Courses Given in Yorkshire by William Smith and John Phillips, by J. M. Edwards. London; W. Arding (1820-25)

Said strata were sometimes separated or reduced by the absence of once present strata but the order in which they were deposited never varied
AD 50-1; OLD RHG 240-1

Described and named formations based on their stratigraphy and lithology and fossils
AD 50-1; CBS 218

VENETZ, IGNACE (1788-1859)

In Journal des Mines, Volume 2 (Read to Société Helvétique des Sciences Naturelles, Berne Meeting)

Said that a glacier had carried the erratics
FOREL 108

1816/17

JAMESON, ROBERT (1774-1854)

In C.G.B. Daubeny, Notes Taken at Jameson´s Lectures on Natural History 1816-17 with Additional Matter Extracted from Other Sources. Oxford College (Manuscript)

Said strata had been regularly deposited without interruption throughout time
OLD RHG 242

1817

BEUDANT, FRANCOIS SULPICE (1787-1850)

Lettre au Sujet du Mémoire de M. Wollaston, in Annales de Chimie, 7: 399-405

Said crystals were result of chemical mixtures in definite proportions
BJG 120

BROCCHI, GIOVANNI BATTISTA (GIAMBATTISTA) (1772-1826)

Catalogo Ragionato di una Raccolta di Rocce Diposto con Ordine Geografico per Servire alla Geognosia dell´ Italia. Milano; Dall´ Imperiale Regia Stamperia

Said universal sea had deposited fossils on mountains
GORT 511; RAM-2 26

FITTON, WILLIAM HENRY (1780-1861)

Article IV Transactions of the Geological Society, in Edinburgh Review, 29: 70-94

Said the earth´s crust was composed of a series of layers in a set order from which a layer might be missing, but never out of order
OLD RHG 206

SMITH, WILLIAM (1769-1839)

Stratigraphical System of Organized Fossils with Reference to the Specimens of the Original Collection in the British Museum Explaining Their State of Preservation and Their Use in Identifying the British Strata. London; E. Williams

Said fossils unique to various strata had been living steps in an ever-improving line of succession
ALB-2 114; CBS 218; HMPD 341

WERNER, ABRAHAM GOTTLOB (1749-1817)

Abraham Gottlob Werner´s Letztes Mineral-System. Freiberg and Vienna; Graz und Gerlach und C. Gerold

Identified 317 minerals and classified them as 1) earthy, 2) saline, 3) combustible, or 4) metallic
AD 204

1818

BREISLAK, SCIPIONE (SCIPION) (1748-1826)

Institutions Géologiques Traduites du Manuscrit Italien en Francais par P. J. L. Campmas Milan; Jean-Pierre Giegler (1819)

Said that early molten earth produced mineral veins when like materials collected together and hardened
AD 320-1

BUCH, CHRISTIAN LEOPOLD VON (1774-1853)

Ueber die Zusammensetzung der basaltischen Inseln und ueber Erhebungs Kratere, in a Lecture Delivered before Prussian Academy of Science, May 1818. In Leonhard´s Taschenbuch für die gesammte Mineralogie, 15: 391-427

Said valleys were created by upward thrust of rocks coincident with mountain building
AD 383

EATON, AMOS (1776-1842)

Index to the Geology of the Northern States. Leicester, MA; H. Brown

Said early earth was a ball of mud in which metals sank to form a core surrounded by granite and shells of strata which cracked and uplifted to form continents when the interior of the earth heated up
FENT 145

Said earth´s crust consisted of 1) primitive nonfossiliferous rocks, 2) transition rocks 3) secondary rocks, 4) superincumbent rocks, and 5) alluvial rocks
FOHYAG 57-9

FITTON, WILLIAM HENRY (1780-1861)

Article III-1 A Delineation of the Strata of England and Wales, Etc., in Edinburgh Review, 29: 310-37

Said strata in several locations on the earth were structurally analogous but were not the same
OLD RHG 242

KAIN, JOHN HENRY (?-1849)

Remarks on the Mineralogy and Geology of the Northwestern Part of the State of Virginia and the Eastern Part of the State of Tennessee, in American Journal of Science, 1: 60-67

Said the French Broad River had been created by a natural catastrophe
GREG 107

MACLURE, WILLIAM (1763-1840)

Essay on the Formation of Rocks, or an Inquiry into the Probable Origin of Their Present Form and Structure, in Academy of Natural Sciences of Philadelphia Journal, 1: 261-276, 285-310, 327-45

Said rocks formed in separate basins and were different so they would not correlate world-wide
MBC 257

Classified rocks as 1) water-deposited, 2) volcanic, and 3) borderline, i.e., of indeterminate origin, and also subdivided all three
HAZ 225-6, 243+

Said the earth was very old, if not ageless
MBC 258

MITCHILL, SAMUEL LATHAM (1764-1831)

Observations on the Geology of North America, in Cuvier´s Essay on the Theory of the Earth: 319-431. New York; Kirk and Mercein

Said the Great Lakes had been a salt sea which broke through the Appalachian barrier and drained eastward, carrying Canadian and United States topsoil with it
CHOR 239; FOHYAG 616-7

WAHLENBERG, GORAN (GEORG) (1780-1851)

Om Svenska Jordens Bildning, in Svea; Tidskrift för Vetenskap Och Konst: 1-99. Uppsala; N.P. (1824)

Said large bodies of water were exits for magnetic forces so petrification of fossils was concentrated under masses of water
PPHG 6

1818-22

SAINT-HILAIRE, ETIENNE GEOFFROY (1772-1844)

Philosophie Anatomique. Paris; J. B. Baillière

Said that all of nature is constantly changing, especially species which react to environmental changes
RUD 151-2

1819

BUCKLAND, WILLIAM (1784-1856)

Vindiciae Geologicae, in The Connexion of Geology with Religion Explained in an Inaugural Lecture Delivered Before the University of Oxford, May 15, 1819. Oxford; University Press (1820)

Said ocean waters evaporated, rose, and returned as rain to feed the springs and rivers
CHOR 109-112

Said strata formed underwater according to the laws of gravitation
CHOR 109-112

GREENOUGH, GEORGE BELLAS (1778-1855)

A Critical Examination of the First Principles of Geology, in Annalen der Physik und Chemie, 14: 365-73, 456-64

Said the earth had been covered by a universal, but temporary, deluge
CHOR 156, 293

MACLURE, WILLIAM (1763-1840)

Hints on Some Outlines of Geological Arrangement with Particular Preference to the System of Werner, in American Journal of Science, 1: 209-13

Said rocks varied in composition from deposit to deposit
MBC 257

MITSCHERLICH, EILHARD (1794-1863)

Ueber die Kristallisation der Salze, in Denen des Metal der Basis mit zwei Proportionen Sauerstoff verbunden ist, in Akademie der Wissenschaften, Berlin. Abhandlungen 19: 427-37

Said a given crystal shape was independent of the chemical nature of its constituent atoms but the number of atoms was definitive
BJG 123

SAY, THOMAS (1787-1834)

Observations on Some Species of Zoophytes, in American Journal of Science, 1: 381-87 and 2: 34-45

Said fossils had chronogenetic value
SCHU 61

1820

BOUE, AMI (1794-1881)

Essai Géologique sur l´Ecosse. Paris; V. Courcier

Said gases and vapors were important for deep seated metamorphic activity
HTOD 42

HAYDEN, HORACE HANDEL (1769-1844)

Geological Essays. Baltimore; J. Robinson

Said alluvial deposits were deposited by a flood which occurred during rapid melting of the polar ice cap
MHG 199-200

SCHLOTHEIM, ERNST FRIEDRICH VON (1764-1832)

Die Petrefactenkunde. Gotha; Becker

Classified fossils by a binomial system
WOOD 70

1821

ESCHSCHOLTZ, JOHANN FRIEDERICH (1793-1831)

Ueber die Korallen-Inseln (On Coral Islands), in Entdeckungreise in die Sudsee und nach der Berings Strasse, 3: Appendix by Otto von Kotzebue. Weimar; von den Gebrudern Hoffmann

Said coral atolls were built atop submarine mountains
STODD 200

HAYDEN, HORACE HANDEL (1769-1844)

Geological Essays (A Review by Silliman), in American Journal of Science, 3: 47-57

Said that drift was distributed by a flood of water
GREG 115

MACCULLOCH, JOHN (1773-1835)

Geological Classification of Rocks, with Descriptive Synopses, of the Species and Varieties Comprising the Elements of Practical Geology. London; Longman, Hurst, Rees, Orme and Brown

Classified rocks, according to geological relationships, as 1) primary, 2) secondary, 3) occasional, and 4) appendix rocks (volcanic, clay, peat, etc.)
CROSS 345-7

MITSCHERLICH, EILHARD (1794-1863)

Om Förhållandet Emellan Chemiska Sammansättningen och Krystall-Formen Hos Arseniksyrade och Phosphorsyrade Salter, in Annales de Chimie, 19: 350-419

Said the same crystal would always result from the same number of atoms which combined in the same manner
SMAJ 18

SILLIMAN, BENJAMIN (1779-1864)

Notice of Horace H. Hayden's Geological Essays, in American Journal of Science, 3: 49

Said that the caverns of the earth had discharged their waters thereby scattering the erratics across the land -- and might do so again
GREG 115

WILSON, J. W.

Bursting of Lakes through Mountains, in American Journal of Science, 3 (2): 252-3

Said the earth's topography had been formed during creation, or the flood, and had changed very little since then
GREG 104

1821

VENETZ, IGNACE (1788-1859)

Mémoire sur les Variations de la Température dans les Alpes de la Suisse, in Société Helvétique des Sciences Naturelles Mémoire, Volume 12 (1833)

Said a great ice sheet had once covered a large area and alpine glaciers and erratics are remnants of it
ALB-2 157; CHOR 193; D&B 425; GIA 160

Said moraines marked advances and retreat of glaciers
GIA 160

1822

BOUE, AMI (1794-1881)

Mémoire Géologique sur l´Allemagne, in Journal de Physique, 94: 297-312, 345-379 and 95: 31-48, 88-112

Said metalliferous veins were result of igneous and aqueous action
HTOD 42

CONYBEARE, WILLIAM DANIEL AND WILLIAM PHILLIPS (1787-1857 / 1773-1828)

Outlines of the Geography of England and Wales. London; W. Phillips

Divided rock masses into formations consisting of series of repeating strata
CONK 40

Said there was an orderly superpositioning of formations and that rocks appeared similar through time with differences mainly caused by age
WOOD 83-4

Said sea and land had altered positions in response to upheavals, depression, reduction in water mass and heightened seafloor resulting from deposition
WOOD 83-4

Said folded, overturned strata was result of a great disturbance
OLD RHG 242

Said fossils were less like modern forms as they increased in age and there was mor similarity of fossils from country to country ages ago than there is now
WOOD 83-4

DAVY, HUMPHRY (1778-1829)

In Collected Works of Sir Humphry Davy, Volumes 1-9, edited by John Davy. London; Smith, Elder and Company (1839-40)

Said primitive life began after the flood with a succession of species and culminated with the rise of man
SEIG 225-6

HAUY, RENE JUST (1743-1822)

Traité de Minéralogie. Paris; Chez Louis

Said rocks were divided into three orders: 1) with visible minerals; 2) without visible minerals; or 3) fragmental
BJG 111; CIR 43; HTOD 144; WMC 69

MOHS, FREDERICK (1773-1839)

Grund-riss der Mineralogie. Dresden; Arnold

Classified crystals as having 1) three or 2) four axes, some of which were mutually perpendicular
SMAJ 19

SCHMIDT, JOHANNES CHRISTIANUS LUDOVICUS (1754-?)

Darstellung mehrerer allgemeiner Verhältnisse der Gange und der Beziehung derselben zur Formazion des Gebirggesteins (Presentation of Several General Relationships of Veins and of Their Connection to the Formation of Mountains), in Karsten's Archiv für Bergbau und Hüttenwesen, 4: 1-61. Berlin; Reimer

Said mineral-rich waters rose from the interior through fissures in which they deposited the minerals in response to cooling
HTOD 41

1823

BUCKLAND, WILLIAM (1784-1856)

Reliquiae Diluvianae (Relics of the Deluge). London; J. Murray

Said rivers could not carve their valleys, therefore, it must have been done by a flood
CHOR 109-18; GREG 107-8

Said resistance of subsurface would affect topography
RUD 136

Said the flood had been world-wide but that human remains found with extinct species were not themselves pre-flood in origin
HMPD 342

Said erratics were transported by a flood
CHOR 109-18; GREG 107-8; GIA 159

Said sandy clay boulder strewn deposits were recent in origin and were 1) alluvial due to ongoing erosion, for example, or 2) diluvial as a result of a current of water from the north
EHGT 136

Said that the flood had carved the river valleys which some rivers were already filling in
GREG 107-8

FLEMING, JOHN (1785-1857)

Review of Cuvier's Essay on the Theory of the Earth (Fourth Edition), in New Edinburgh Review, 4: 383-4, 387-95

Said remains of extinct and extant specie in the same strata was a result of extinction being piece-meal
WDC 270; RUD 172

GAY-LUSSAC, JOSEPH LOUIS (LOUIS JOSEPH) (1778-1850)

Réflexions sur les Volcans, in Annales de Chimie, 22: 415-29

Said earthquakes were waves of vibration
DAV 68

HITCHCOCK, EDWARD (1793-1864)

A Sketch of the Geology, Mineralogy, and Scenery of the Regions Contiguous to the River Connecticut, in American Journal of Science, 7: 1-30

Said that the gorge of the lower Connecticut River had been formed when a breach drained several lakes
GREG 107

HUMBOLDT, FRIEDERICH HEINRICH ALEXANDER VON (1769-1859)

Essai Géognostique sur le Gisement des Roches dans les Deux Hémisphères (A Geognostic Essay on the Superposition of Rocks). Paris; F. G. Levrault

Said rock layers occurred in a set sequence: granite, gneiss, mica-slate, ophiolite, syenite, porphyry, calcareous and gypseous, elastics, coal, and rock-sa
OLD RHG 206

Said sedimentary rock had the same composition world-wide but that the relative percentages of components might differ between formations
OLD RHG 204-5; VON 32-3; WOOD 80

Said strata were parallel thus causing a given strike, and sometimes the dip, to be uniform region-wide
VON 31

1823-25

LEONHARD, KARL CAESAR VON (1779-1862)

Charakteristik der Felsarten. Heidelberg; J. Engelmann

Classified rocks as 1) heterogeneous, 2) homogeneous, 3) fragmented, 4) loose, or 5) other (coal, etc.)
CROSS 347-8

Said metal deposits were classified separately from rocks
HTOD 141

1824

BEAUMONT, LEONCE ELIE DE (1798-1874)

Faits pour Servir à l'Histoire des Montagnes de l'Oisans, in Annales des Mines, Third Series, 1834: 3-63

Said topography of Alps since dislocation was unchanged by erosion
MARG 103

ESMARK, JENS (1763-1839)

Bidrag Til Vor Jordklodes Historie, in Magazin for Naturvidenskaberne, 3: 28-49

Said glacier had covered the continent
D&B 425

HITCHCOCK, EDWARD (1793-1864)

Review of Buckland's Reliquiae Diluvianae, in American Journal of Science, 7: 332

Said all diluvial phenomena were accounted for by avalanches and flooding by the waters present in the streams and lakes
WDC 266

MOHS, FREDERICK (1773-1839)

Grund-riss der Mineralogie. Dresden; in der Arnoldischen Buchhandlung (1822-24)

Arranged ten common minerals according to hardness and measured all others against them
HURL 48

SILLIMAN, BENJAMIN (1779-1864)

Review of "Outlines of the Geology of England and Wales ... by W.D. Conybeare and Wm. Phillips", in American Journal of Science, 7: 238

Said some trap and granite were igneous in origin
GKES 142

WEBSTER, THOMAS (1773-1844)

Observations on the Purbeck and Portland Beds, in Geological Society of London Transactions, 2: 37-44 (1829)

Said the strata contained alternating layers of marine and fresh-water strata
GEI 397

WHEWELL, WILLIAM (1794-1866)

General Method of Calculating the Angles Made by Any Planes of Crystals and the Laws According to Which They Are Formed, in Royal Society of London. Philosophical Transactions, 115: 87-130	Used coordinate geometry to describe crystals DEAS 135

1825

BUCH, CHRISTIAN LEOPOLD VON (1774-1853)

Physicalische Beschreibung der Canarischen Inseln. Berlin; Königliche Akademie der Wissenschaften	Said gravity caused differentiation to occur after which crystallization took place HSP 75
	Said mountain building occurred when 1) rising magma arches crust over to form a dome mountain or 2) breaks through to form a volcano AD 381-6; VON 31; TIK 324

DOBSON, PETER

Remarks on Bowlders, in American Journal of Science, 10: 217-8 (1826)	Said local boulders were bottom-smooth, worn by being dragged over rocks and gravel by ice CHOR 193-4, 292; FAIR 155; HLF 155-6; MERR 273
	Said erratics were carried by ice, under water GREG 116

LEA, ISAAC (1792-1886)

On Earthquakes, Their Causes and Effects, in American Journal of Science, 9: 204-15	Said earth´s incandescent interior contained tunnels which connected volcanoes and contained explosive gases and the rumblings of earthquakes MERR 280

RENSSELAER, JEREMIAH VAN (1793-1871)

Lectures on Geology. New York; E. Bliss and E. White	Said successive generations of fossil remains, differing more as increasing distance separated the deposits, identified geological horizons FOHYAG 107

SCROPE, GEORGE JULIUS DUNCOMBE POULETT (1797-1876)

Considerations on Volcanoes, the Probable Causes of Their Phenomena, the Laws Which Determine Their March, the Disposition of Their Products, and Their Connexion with the Present State and Past History of the Globe; Leading to the Establishment of a New Abstract of a Dissertation ... Concerning the System of the Earth, Its Duration and Stability. London; W. Phillips

Said all types of igneous rock formed from one parent magma after liquation and differentiation prior to crystallization of the clay-and-water-like mixture
HSP 61, 68, 72

Said earthquakes preceded volcanic activity, decreasing when eruption began
DAV 155, 228

Said extinct volcanoes were no different from active volcanoes, except for their quiescence
RUD 170

Said the steps involved in valley cutting were discernible in lava flows
RUD 170

Said volcanoes were situated in a linear fashion, one after another
EOSG 276

Said water and heat in lava resulted in fluidity and volatiles were also an important component
HTOD 42; HSP 73

SEDGWICK, ADAM (1785-1873)

Origin of Alluvial and Diluvial Formations, in Annals of Philosophy, 9: 241-57 and 10: 18-37

Said fragmented, irregularly distributed rocks of unusual composition had been deposited by flood waters but other deposits had been river transported
ALB-2 144

1826

DAUBENY, CHARLES GILES BRIDLE (1795-1867)

A Description of Active and Extinct Volcanoes. London; W. Phillips

Said earthquakes and volcanoes resulted from heat released when subsurface water mixed with uncombined bases beneath the crust
WOOD 81

EMMONS, EBENEZER (1799-1863)

Manual of Mineralogy and Geology. Albany; Websters and Skinners

Classified minerals as 1) oxidable, 2) metallic, 3) alkaline, 4) an earth or compound of 4) and 2) or 3)
FOHYAG 110

1826

GEDDES, JAMES (1763-1838)

Observations on the Geological Features of the South Side of Ontario Valley, in American Journal of Science, 11: 213-18

Said valleys predated their waters
GREG 109

1827

BRONGNIART, ALEXANDRE (1770-1847)

Classification et Caractères Minéralogiques des Roches Homogènes et Hétérogènes. Paris; F. G. Levrault

Classified rocks by visible mineral components: 1) homogeneous, or 2) heterogeneous, the latter divided into a) of crystals, and b) aggregate
CIR 44; CROSS 350-1; GRA 39; HTOD 156-7

Defined structure as large-scale features, and texture as fine features, of rock
HSP 28

FITTON, WILLIAM HENRY (1780-1861)

An Account of Some Geological Specimens, in Narrative of a Survey of the Intertropical and Western Coasts of Australia, 1818-22, by Phillip Parker King. London; J. Murray

Said limestones have been forming throughout time and formations could be dated by their fossils
SEDD 393

NICOL, WILLIAM (1768?-1851)

Neue Methodie die Structur fossiler Vegetabilien zu untersuchen, in Froriep´s Notizen, 30: 97-9 (1831) (see Observations on Fossil Vegetables by Henry Witham)

Studied fossils by examning thin sections under microscope and comparing them with recent plants
GEI 463-4

VANUXEM, LARDNER (1792-1848)

Proofs Drawn from Geology of the Abstraction of Nitrogen from the Atmosphere by Organization, in Silliman Journal (American Journal of Science), 12: 84-93

Said the earth´s crust interacted chemically with the atmosphere
FAIR 4; HER 42

1828

BRONGNIART, ADOLPHE THEOPHILE (1801-1876)

Histoire des Végétaux Fossiles, ou Recherches Botaniques et Géologiques sur les Végétaux Renfermés dans les Diverses Couches du Globe. Paris; G. Dufour	Said local climate had once been as hot as the tropics and the earth was cooling down RUD 147
Prodrome d'une Histoire des Végétaux Fossils (Forerunner of a History of Fossil Plants). Paris; F. G. Levrault	Said there were four periods of gradual change, divided by discontinuities, in development of plant life: 1) domination by vascular cryptograms, 2) appearance of conifers, 3) appearance of cycads, and 4) first dicotyledons RUD 146
Notice sur les Blocs de Roches des Terrains de Transport en Suède, in Annales des Sciences Naturelles, Series 1, 14: 1-22	Said the erratics were carried by great ocean currents to their present locations GIA 159

EGEN, PETER NIKOLAUS CASPER (1793-1849)

Ueber das Erdbeben in den Rhein- und Niederlanden vom 23 Febr. 1828, in Annalen der Physik und Chemie, 13: 153-63, 176-9	Formulated a six degree scale for measuring earthquakes DAV 40

LESLIE, JOHN (1766-1832)

Theory of Compression Applied to Discover the Internal Constitution of Our Earth, in Edinburgh New Philosophical Journal, 6: 84-9	Said the earth was filled with light which would be unaffected by the pressure exerted there NCDAFG 231

MITCHELL, ELISHA (1793-1857)

On the Character and Origin of the Low Country of North Carolina, in Silliman Journal (American Journal of Science), 13: 336-48	Said strata had formed in the ocean and had been forced upward before the Noachian flood FOHYAG 115

NICOL, WILLIAM (1768?-1851)

Observations on the Fluids Contained in Crystallised Minerals, in Edinburgh New Philosophical Journal, 5: 94-6	Studied the behavior of light in crystals by use of his polarizing microscope DANA 6

PHILLIPS, JOHN (1800-74)

Remarks on the Geology on the North Side of the Vale of Pickering, in Philosophical Magazine, New Series, 3: 243-9

Said an escarpment formed from the effect of erosion on tilted strata
CHOR 134

Said erosion by rain and rivers destroyed hills and mountains and filled valleys with the debris
CHOR 134

Said the erratics had been deposited during the flood
CHOR 134

1828?

DAVY, HUMPHRY (1778-1829)

In Collected Works of Sir Humphry Davy, Volumes 1-9 edited by John Davy. London; Smith, Elder and Company (1839-40)

Said volcanoes resulted from the earth´s central fire
NCDAFG 230

1829

BEAUMONT, LEONCE ELIE DE (1798-1874)

Recherches sur Quelques-Unes des Révolutions de la Surface du Globe, in Revue Francaise, 15 (May 1830) and Annales des Sciences Naturelles, 18: 5-25, 284-416 (1829)

Said mountain uplift was the result of horizontal tangential forces produced by crustal shrinking in response to cooling of the earth´s interior
GER 27, 33

Said mountain ranges came into existence suddenly and were catastrophic in origin
AD 394

BECHE, HENRY THOMAS DE LA (1796-1855)

Notice on the Excavation of Valleys, in Philosophical Magazine, New Series, 6: 241-8

Said valleys were formed by violent catastrophes or floods (rounded outlines) or by rivers (narrow gorges) and that dry valleys had been flood created
CHOR 130-3, 292

CONYBEARE, WILLIAM DANIEL (1787-1857)

On the Hydrographical Basin of the Thames, in Geological Society of London Proceedings, 1: 145-9

Said river beds do not change so debris at the mouth of the river had to be the result of heavy floods
CHOR 119-24, 292

Said the depth and steepness of valleys were result of some violent force but dry valleys had been carved by the flood
CHOR 119-24, 292

ACLURE, WILLIAM (1763-1840)

Remarks on the Igneous Theory -- Abstract of a Dissertation ... Concerning the System of the Earth, Its Duration and Stability, in Silliman Journal (American Journal of Science), 16: 351-2

Said masses of fermenting organic material could release elastic gases and thus cause earthquakes
MERR 280

Remarks on the Theory of a Central Heat in the Earth, and on Other Geological Theories, in American Journal of Science, 15: 384-6

Said the primitive rocks were igneous but once were sedimentary formations subsequently heated to a molten state before recrystallizing
MBC 251, 256

HILLIPS, JOHN (1800-74)

Organic Remains of the Eastern Part of Yorkshire, in His Illustrations of the Geology of Yorkshire: 111-18. York; T. Wilson and Sons

Grouped strata in a formation by the presence of fossil types found above and below
CONK 76, 204

Said strata could be correlated by comparing their order, fossils and mineral composition
CONK 75

EDGWICK, ADAM (1785-1873)

On the Geological Relations and Internal Structure of the Magnesian Limestone, in Geological Society of London Transactions, Second Series: iii

Said unconformities sometimes merged with others, or did not reappear at all so they could not be traced worldwide
CONK 76

ILLIMAN, BENJAMIN (1779-1864)

Outline of the Course of Geological Lectures Given in Yale College. New Haven; Hezekiah Howe

Said waters covering early earth receded into the interior, after the mountains formed, but a universal deluge subsequently resubmerged the land, causing extinction of animal life
MERR 292-4

URE, ANDREW (1778-1857)

A New System of Geology in Which the Great Revolutions of the Earth and Animated Nature Are Reconciled at Once To Modern Science and Sacred History. London; Longman, Rees, Orme, Brown and Green

Said a great convulsion had sunk the continent which now lay beneath the Pacific Ocean
STODD 202

VANUXEM, LARDNER (1792-1848)

Remarks on the Characters and Classification of Certain American Rock Formations, in Silliman Journal (American Journal of Science), 16: 25-6

Said rocks could be unequivocally identified by 1) their fossils, then 2) position and 3) mineral content
FOHYAG 117-8, 122

1829-30

BEAUMONT, LEONCE ELIE DE (1798-1874)

Recherches sur Quelques-Unes des Révolutions de la Surface du Globe, in Annales des Sciences Naturelles, 18: 5-25, 284-416, and 19: 177-240

Said that mountain building had occurred during different periods of the earth's history in response to sudden release of internal stresses
RUD 142

Said episodes of mountain building coincided with major formal discontinuities which could be dated by comparing dates of strata on either side of the discontinuity, and thus respective ages of mountains could be determined
MARG 103; RUD 142

Said all mountain ranges of the same age were parallel or cut the meridian at the same angle and each belonged to one system, the intersections of which formed a pentagonal network
AD 394

Said as earth cooled from hot gaseous ball and contracted, it cracked and parts rose to form mountain ranges
AD 393-4; BRUSH 707; NCDAFG 229

SCROPE, GEORGE JULIUS DUNCOMBE POULETT (1797-1876)

On the Gradual Excavation of the Valleys in Which the Meuse, the Moselle, and Some Other Rivers Flow, in Geological Society of London Proceedings, 1: 170-1

Said rivers cut valleys and slow moving streams had cut the meanders which occur in resistant rock
CHOR 129, 296; GREG 108

1830

EATON, AMOS (1776-1842)

Geological Textbook. Albany; Websters and Skinners

Said strata are arranged in five series: 1) volcanic, 2) diluvian, 3) post diluvian, 4) analluvian, 5) anomalous and each of these series consists of 1) carboniferous (lowest), then 2) quartzose, and 3) calcareous formations
FOHYAG 128-31

Said earth was originally layered concentrically, and when an inner force broke the rings, the debris rose to the surface
FOHYAG 128-30

HESSEL, JOHANN FRIEDRICH CHRISTIAN (1796-1872)

Kristallometrie. Leipzig; E. B. Schwickert

Said thirty-two types of symmetry were possible in crystals
RPM 13-4

HUGI, FRIEDRICH JOSEPH (1796-1855)

Naturhistorische Alpenreise. Solothurn; Amiet-Lutiger

Said glacial movement was effected by its internal workings
GIA 158

LYELL, CHARLES (1797-1875)

Letter to G. A. Montell, 15 February, in The Life Letters and Journals of Sir Charles Lyell, Bart. London; John Murray (1881)

Said conditions on earth varied from a mean so if conditions of ancient times returned, so might the extinct organisms of that time
OLD RHG 245

MACLURE, WILLIAM (1763-1840)

Geological Remarks Relating to Mexico, in American Journal of Science, 20: 406-8 (1831)

Said veins were filled by vapors ascending, through cracks, from lava
MBC 251, 256

YATES, JAMES (1789-1871)

On the Formation of Alluvial Deposits, in Geological Society of London Proceedings, 1: 237-9 (1826-33)

Said alluvium was formed from materials supplied by earthquakes, landslides, frost and oxidation, and by running water which rounds and separates sedimentary particles by weight
CHOR 283

1830-33

LYELL, CHARLES (1797-1875)

Principles of Geology. London; J. Murray

Said abrupt changes between strata occur when deposition ceases for a period of time
SLPG 24

Said all species, except the recent arrival, man, had appeared, fully developed, during creation which is a recurring, but non-progressive event as is the extinction of species
HUBB 236; CHA 339; RUD 182; SLPG 18-21

Said climate fluctuates around a mean in respose to the influences of land or sea wind and current directions, etc., so exchange of land and sea affected ancient temperatures and present cold region may have been tropical
RUD 180-1; SLPG 13

Said erosion by the sea retarded the elevation of land which occurred when deep-seated igneous activity caused earthquakes
CHOR 307; SLPG 16

Said erratics were deposited by icebergs which had split from land glaciers and floated away carrying the debris
CHOR 164; SLPG 28

Said glaciers occur when the weight of accumulated snow forces it down-valley carrying detritus which it drops off in long ridges
CHOR 164; SLPG 28

Said interior of earth was equally hot throughout time
NCDAFG 230

Said metamorphism accounted for the lack of fossils in the early primary rocks
RUD 184

Said petroleum was distilled from buried organic materials by earth´s inner fires
OWEN 30

Said some erratics were carried by icebergs and some by waters of glacier-
dammed lakes freed by earthquakes
GIA 159

Said species arise and die out at various
times and places for varied periods but
overall, population varies uniformly so
proportion of fossils in different strata
date them
CONK 78; SLPG 24-5

Said streams could carry large particles
because the material weighed less in
water
CHOR 160

Said that aqueous and volcanic activity
builds as well as destroys, with a
balance being maintained between growth
and destruction
SLPG 15, 17

Said that while Europe and North America
were water-covered, icebergs deposited
the erratics
CHOR 200; EHGT 137

Said the changes which occurred in
the past were the same ones occurring
today and tomorrow throughout time,
gradually and invariably repeating the
cycle of uplift, erosion, return of the
ocean, etc.
ALB-2 230, 234, 251; BEL 32-3; CLOUD 50;
GNC 25

Said the present distribution of plants
and animals developed in response to
environmental, especially climatic,
changes in their respective centers of
origin
STODD 211

Said traces of earth´s past history are
present in the rocks and can be applied
to present day happenings
GNC 25; OLD RHG 244

1831

BECHE, HENRY THOMAS DE LA (1796-1855)

Geological Manual. London; Treuttel and Wurtz, Treuttel, Jun and Richter

Said ice was a possible explanation for the transport of erratics
CGT 17; CHOR 193-4

BRONN, HEINRICH GEORG (1800-62)

Italiens Tertiär-Gebilde und deren organische Einsschlüsse (Vier Abhandlungen). Heidelberg; Buchandlung von K. Groos

Dated deposits and units of time relatively by calculating relative numbers of mollusc species present
RUD 190

DESHAYES, GERARD-PAUL (1795-1875)

Tableau Comparatif des Espèces de Coquilles Vivantes avec les Espèces de Coquilles Fossiles des Terrains Tertiaires de l´Europe, et des Espèces de Fossiles de ces Terrains Entr´eux, in Société Géologique de France Bulletin, 1: 185-7

Said deposits and units of time could be dated relatively by calculating the relative proportions of species of molluscs present
RUD 190

PHILLIPS, JOHN (1800-74)

On Some Effects of the Atmosphere in Wasting the Surfaces of Buildings and Rocks, in Geological Society of London Proceedings, 1: 323-4

Said atmospheric influences caused exfoliation
CHOR 289

Said the pattern caused by weathering was a result of the rock´s composition
CHOR 289

Said weathering was a rapid wasting caused by alternating hot/cold and dry/wet conditions
CHOR 289

TRIMMER, JOSHUA (1795-1857)

On the Diluvial Deposits of Caernarvonshire, in Geological Society of London Proceedings, 1: 331-2

Said deposits of gravel, sometimes with shell fragments, sand and clay had been carried by flood waters
CGT 10

WITHAM, HENRY THORNTON MAIRE (1779-1844)

Description of a Fossil Tree Discovered in the Quarry of Craigleith, in Natural History Society of Northumberland, Durham and Newcastle-upon-Tyne Transaction, 1: 294-301

Described fossils by studying them in thin section
YODER 8

ZEMBNITSKY, Y.G.

Conchyliology, or Presentation of Data on Shells and Animals That Produce Them, Volume Eight. St. Petersburg; N.P.

Said the external features of fossils were an expression of the environment in which they had lived
TDG 381

1832

BERNHARDI, REINHARD (FL. 1830-41)

Wie Kamen die aus dem norden stammenden Felsbruchstücke und Geschiebe, in Jahrbuch für Mineralogie, 3: 257-67

Said polar ice had reached as far as there were erratics to be found, leaving them behind when it melted
CGT 20; CHOR 193-4; MHG 200

PHILLIPS, JOHN (1800-74)

Geology of Yorkshire, in American Journal of Science, 21: 17-20

Said that only a great flood could have cut the river valleys
GREG 108

SAUSSURE, LOUIS ALBERT NECKER DE (1786-1861)

An Attempt to Bring under General Geological Laws the Relative Position of Metalliferous Deposits, with Regard to the Rock Formations of Which the Crust of the Earth is Formed, in Geological Society of London Proceedings, 1: 392-4 (1834)

Said igneous fire affecting crystalline rocks produced metalliferous veins by sublimation
HTOD 44-61; TAY 22

SMITH, ALFRED (1822?-98)

On the River Courses and the Alluvial and Rock Formations of the Connecticut River Valley, in American Journal of Science, 22 (2): 205-31

Said varved deposits had formed in rock-dammed lakes
CHOR 287

1833

AGASSIZ, LOUIS (1807-73)

Recherches sur les Poissons Fossiles (On Fossil Fishes). Neuchâtel (Suisse); Imprimerie de Petitpierre

Said that the successive changes in fish recorded in the fossil record correspond[e] with those of the developing embryo
HMPD 346

AMPERE, ANDRE-MARIE (1775-1836)

Théorie de la Terre d'Après M. Ampère, in Revue des Deux Mondes, 3: 106-7

Said earth was solid throughout its interior
BRUSH 707

Said the earth cooled in layers around the solid core and chemical reaction between the layers caused volcanoes
NCDAFG 229

BECHE, HENRY THOMAS DE LA (1796-1855)

Geological Manual. London; Charles Knight

Classified rocks as 1) stratified (superior-fossiliferous or inferior-nonfossiliferous) or 2) unstratified
CHOR 193-4

HITCHCOCK, EDWARD (1793-1864)

Final Report on the Geology of Massachusetts. Northampton; J. H. Butler (1841)

Said granite was igneous and had been forced into sedimentary rocks overlying its molten origin in the earth's hot interior
GKES 143

Said rows of overturned rocks has formed on an incline or had been eroded on a slant
GKES 152-3

MURCHISON, RODERICK IMPEY (1792-1871)

On the Sedimentary Deposits Which Occupy the Western Parts of Shropshire and Herefordshire and Are Prolonged from NE to SW through Radnor, Brechknock, and Caermarthenshires, with Descriptions of the Accompanying Rocks of Intrusive or Igneous Characters, in Geological Society of London Proceedings, 1: 470-7

Said fossiliferous Silurian strata suggested a warmer climate had prevailed world-wide
CONK 81; RUD 191-4

Said transitional nature of worldwide strata between unfossiliferous secondary zones suggested many plants and animals had not appeared yet
RUD 191-4

1833-34

GREENOUGH, GEORGE BELLAS (1778-1855)

Anniversary Address, in Geological Society of London Proceedings, 2 (35): 42-70

Said that even if there had been a deluge, there would be no evidence left
CHOR 156-7, 293

1834

BABBAGE, CHARLES (1792-1871)

Observations on the Temple of Serapis, at Pozzuoli, Near Naples, with Remarks on Certain Causes Which May Produce Geological Cycles of Great Extent, in Geological Society of London Quarterly Journal, 3: 186-217 (1847)

Suggested that earthquakes resulted when heated gases in the earth cause the crust to crack
GER 34

Suggested that there was molten lava beneath the earth´s crust
GER 34

Suggested that mountain uplift was effected by earthquakes
GER 34

CHARPENTIER, JOHANN VON (JEAN G. F. DE) (1786-1855)

Notice sur la Cause Probable de Transport des Blocs Erratiques de la Suisse, in Annales des Mines, 8: 219-36 (1835)

Said alpine glaciation had been wide-spread when Alps were higher
ALB-2 157; GIA 160

Said glaciers moved when water froze in fissures and expanded
CGT 17-8; ROWL 190

Said erratics were carried by ice
CGT 17-8; CHOR 194

D´AOUST, THEODORE VIRLET (1800-94)

Nouvelles Notes Relatives à l´Origine Volcaniques des Bitumes Minéraux, in Société Géologique de France Bulletin, 4: 372-76

Said petroleum was a nonorganic, volcanic product
OWEN 31

FOURNET, JOSEPH JEAN BAPTISTE XAVIER (1801-69)

Etudes sur les Dépôts Métallifères, in D´Aubuisson de Voisins´ Traité de Géognosie, Volume III, 1835. Paris; F.G. Levrault

Said ascending waters and igneous vapors contributed to metal ore deposits
HTOD 49

LYELL, CHARLES (1797-1875)

On the Proofs of a Gradual Rising of the Land in Certain Parts of Sweden, in *Royal Society of London. Philosophical Transactions*, 125: 1-38

Said elevation of mountains and land could occur without earthquakes
SLPG 16

MERIAN, PETER (1795-1883)

Ueber die in Basel wahrgenommenen Erdbeben. Basel; A. Wieland

Said earthquakes occurred oftener in the winter
DAV 38-9

MOHL, HUGO VON (1805-1872)

Ueber den Bau und die Formen von Pollenkorner. Bern; Fischer

Detailed a descriptive classification of pollen
HCMP 279

REICHENBACH, KARL LUDWIG FRIEDERICH (1788-1869)

On Petroleum or Mineral Oil, in *Journal fur Chemie und Physik (Schweigger´s Journal)*, 69 (Jahrbuch 9): 133 (Translation in *Edinburgh New Philosophical Journal*, 16: 376-84)

Said petroleum was not a product of coal
OWEN 32

1835

DARWIN, CHARLES ROBERT (1809-82)

Observations of Proofs of Recent Elevation on the Coast of Chili, in *Geological Society of London Proceedings*, 2: 446-9 (1837)

Said the Chilean coast was elevated by an 1835 earthquake
RUD 189

Structure and Distribution of Coral Reefs, in *Journal of the Royal Geographical Society of London*, 12: 115-9

Said coral atolls occurred during evolution of an island after volcanoes built i and the sea floor subsided, leaving only the coral reef, which continued to grow upward as subsidence continued
D&B 141; STODD 205

FORCHHAMMER, JOHAN GEORG (1794-1865)

Danmarks Geognostiske Forhold. Kjobenhavn; J. H. Schultz

Said all rocks contained some heavy metal deposited there by circulating subsurface waters, which also created veins
AD 319

HOPKINS, WILLIAM (1793-1866)

Researches in Physical Geology, in Royal Society of London. Philosophical Transactions, 7: 1-84 (1838)

Said that subcrustal convection occurred
MEYER 6564

LEONHARD, KARL CAESAR VON (1779-1862)

Lehrbuch der Geognosie und Geologie. Stuttgart; E. Schweizerbart´s Verlagshandlung

Classified rocks as 1) heterogeneous, 2) homogeneous, 3) crystalline, 4) cemented clastics, or 5) uncemented clastics
CIR 44; HTOD 154

POISSON, SIMON DENIS (1781-1840)

Théorie Mathématique de la Chaleur. Paris; Bachelier

Said the core of the earth was solid and internal heat had been absorbed when the planet passed through a hot region in space
NCDAFG 230

SCHIMPER, KARL (1803-1867)

Ueber die Eiszeit, in Actes de la Société Helvétique des Sciences Naturelles: 38-51 (1837)

Said ice fields had once covered Europe and dumped erratics in their wake when they retreated
CGT 19; CHOR 196

SMITH, WILLIAM (1769-1839)

Deductions from Established Facts in Geology. By the Author (a Broadsheet)

Said the earth developed in stages, 1) rocks, no life, 2-4) marine organisms appeared, died, and new forms arose, 5) land appeared, 6) plants and animals developed, 7) the Flood, 8) the Ice Ages, and 9) to present
ALB-2 114-5

1836

BUCKLAND, WILLIAM (1784-1856)

Geology and Mineralogy Considered with Reference to Natural Theology. London; W. Pickering

Said the biblical ´days´ were long periods of time
CHOR 172-3

Said expansive powers of heat and vapour raised the first rocks and continue to uplift rocks and fuel volcanoes
CHOR 172-3

DESMAREST, NICHOLAS (1725-1815)

Extrait d'un Mémoire sur la Détermination de Quelques Epoques de la Nature par les Produits des Volcans, and Sur l'Usage de ces Epoques dans l'Etude des Volcans, in Observations sur la Physique, sur l'Historie Naturelle et sur les Arts, 13: 115-26

Said that the processes at work in nature were the same through time but that volcanoes were accidental
TND 350

FITTON, WILLIAM HENRY (1780-1861)

Observations on Some of the Strata between the Chalk and the Oxford Oolite, in the South-East of England, in Geological Society of London Transactions, Series 2, 4 (2): 103-5, 328-34

Said there were marine deposits contemporaneous with nearby freshwater deposits
CONK 79

FOX, ROBERT WERE (1789-1877)

On the Formation of Mineral Veins, in Geological Society of London Proceedings, 2: 406, 539-40 (1838)

Said minerals were deposited in response to electrical differences between walls of fissures, and the varying electronegative conditions of rocks made minerals more easily attracted by certain rocks
HTOD 54-5; TAY 18

GESNER, ABRAHAM (1797-1864)

Remarks on the Geology and Mineralogy of Nova Scotia. Halifax; Gossip and Coade

Said coal was organic material deposited on the sea floor, and uplifted by volcanic action
MERR 338-9

Said drift was deposited during several universal inundations by the sea but erratics on hilltops were thrown there by volcanic eruptions
MERR 338-9

Said slate formed in warm waters as evidenced by the fossil shells in it
FOHYAG 177

GIBSON, JOHN BANNISTER (1780-1853)

Remarks on the Geology of the Lakes and the Valley of the Mississippi, in <u>American Journal of Science</u>, 29: 201-13

Said the ocean had invaded the region between the Rocky Mountains and the Allegheny thus accounting for water related phenomena
GREG 115

HERSCHEL, JOHN FREDERICK WILLIAM (1792-1871)

Letter, in <u>Geological Society of London Proceedings</u>, 2: 597

Said that plasticity of the crust permits deeply buried rocks to flow outward in response to sediment loading in spite of the forces of pressure
DALY 1-2

Letters Quoted by Charles Babbage, in the <u>Ninth Bridgewater Treatise</u>. London; John Murray (1837)

Said that surface topography paralleled subsurface isothermals which heated thick sediments forcing them surfaceward as an uplift event or a volcano
DRH 260

Said loading by sedimentation increased pressure eventually causing lateral flow of lower layers and subsidence
DRH 259-60

HILDRETH, SAMUEL PRESCOTT (1783-1863)

Observations on the Bituminous Coal Deposits of the Valley of the Ohio, and the Accompanying Strata, in <u>American Journal of Science</u>, 29 (1): 1-148

Said oil and gas were genetically related to coal
OWEN 39

Said some ridges and valleys, in folded strata, antedated their streams
GREG 109

Said West Virginia ridges and valleys formed before water filled them, and the water gaps had earlier been lakes
GREG 109

HUNTON, LOUIS

Accompanying Remarks to a Section of the Upper Lias and Marlstone of Yorkshire, Showing the Limited Vertical Range of the Species of Ammonites, and Other Testacea, with Their Value as Geological Tests, in <u>Geological Society of London Transactions</u>, 5: 215-221 (1840)

Said there were paleontological zones which occurred irrespective of lithology or its changes
CONK 79; WSSB 222-3

1836

JACKSON, CHARLES THOMAS (1805-1880)

First Report on the Geology of the State of Maine. Augusta; Smith and Robinson (1837)

Said gravel ridges had been formed by a strong current during the last great deluge
FOYHAG 622

MURCHISON, RODERICK IMPEY (1792-1871)

The Gravel and Alluvia of S. Wales and Siluria, as Distinguished from a Northern Drift Covering Lancashire, Cheshire, N. Salop, and Parts of Worchester and Gloucester, in *Geological Society of London Proceedings*, 2 (43): 230-336

Said drift deposit erratics could be transported by icebergs
EHGT 137

1837

AGASSIZ, LOUIS (1807-1873)

Discours de Neuchâtel, in *Edinburgh New Philosophical Journal*, 24 (48): 364-83 (1838)

Said all diluvial phenomena were simply glacier-borne erratics
CHOR 196; FOYHAG 623-4

Said northern ice reached south to he Mediterranean Sea before the Alps were uplifted and that uplift, which occurred during the ice-age after a worldwide drop in temperature, broke up the ice and tumbled debris down the glaciers remaining
ALB-2 160; CGT 18; CHOR 196; EHGT 137; GIA 160

Said that glaciers occurred when surface water entered fissures in the ice, froze, expanded and thus caused downslope movement
CHOR 196

Said there has been alternate destruction and creation of species, by abrupt changes in temperature, and during each large-scale loss of life, as collective body-heat of the dead organisms dissipated and earth´s temperature fell, an ice age began
AIGT 8; ALB-2 159-60; CHOR 196; D&B 425; GIA 161

BISCHOF, KARL GUSTAV (1792-1870)

Die Warmelehre des innern unsers Erdkörpers (Physical, Chemical, and Geological Researches on the Internal Heat of the Globe). London; Longman, Orme, Brown, Green and Longmans (1841)

Said active volcanoes were interconnected through the molten inner earth which caused earthquakes
NCDAFG 228

BUCKLAND, WILLIAM (1784-1856)

Geology and Mineralogy, Considered with Reference to Natural Theology. London; W. Pickering

Said the surface of the earth had changed through time in a series of "creative" operations
CHOR 172-3

DANA, JAMES DWIGHT (1813-95)

A System of Mineralogy. New Haven; Durrie and Peck and Herrick and Noyes

Said chemical identification of a species would agree with visual identification if the crystallization were sufficiently clear
GRA 34

DARWIN, CHARLES ROBERT (1809-82)

On Certain Areas of Elevation and Subsidence in the Pacific and Indian Oceans, as Deduced from the Study of Coral Formations, in Geological Society of London Proceedings, 2: 552-4

Said some areas of oceanic crust were subsiding while others were uplifting to form islands
RUD 189

HERSCHEL, JOHN FREDERICK WILLIAM (1792-1871)

In a Letter to Lyell

Suggested loading by sediment along ocean edge caused subsidence of sea floor and uplift of the continental margin to create mountains
AD 397; FAIR 54-5

LYELL, CHARLES (1797-1875)

Principles of Geology (Fifth Edition). London; J. Murray

Said ice surrounded rocks making them lighter and easier to move when disturbed by earthquakes, avalanches, bursting dams, or glaciers
EHGT 137

Said streams responded to increased discharge by deepening their channels
CHOR 160

POISSON, SIMON DENIS (1781-1840)

Mémoire sur les Températures de la Partie Solide du Globe, de l´Atmosphère, et du Lieu de l´Espace où la Terre se Trouve Actuellement, in Académie des Sciences, Paris. Comptes Rendus, 4: 137-66

Said outer portions of a cooling liquid earth would sink to the center as they cooled and contracted, so the interior would be solid by the time the crust formed
BRUSH 707; NCDAFG 230

WHEWELL, WILLIAM (1794-1866)

History of the Inductive Sciences. London; J. W. Parker

Said the earth´s hot inner core, though not necessarily a fluid mass, was responsible for volcanoes, earthquakes and the globe´s spherical shape
NCDAFG 231

WILLIAMS, JOHN (1796-1839)

Narrative of Missionary Enterprises in the South Sea Islands. New York; D. Appleton and Co.

Said cellular corals become crystalline when exposed to an electric field in the ocean
STODD 208

Said there was a class of Pacific island, of coral limestone, which resulted from volcanic action or expansion
STODD 201

1838

BRUNET

La Découverte d´Os d´Eléphant, in Société Géologique de France Bulletin, 9: 252

Said petroleum and bitumens and igneous phenomena result from the same actions
OWEN 31

EHRENBERG, CHRISTIAN GOTTFRIED (1795-1876)

Ueber die Bildung der Kreidefelson und des Kreidemergels durch unsichtbare Organismen, in Akademie der Wissenschaften Abhandlungen, 23: 59-138

Said Cretaceous rocks contained fossils of still living organisms
OWEN 32

GRESSLY, AMANZ (1814-65)

Observations Géologiques sur le Jura Soleurois, in Société Helvétique des Sciences Naturelles Mémoire, 2: 10-12, 20-22. Neuchâtel; Imprimerie de Petitpierre (1838-41)

Decribed lithological and paleontological makeup of units he called facies
CONK 79; HDMF 152; TIK 325, 329

HALL, JAMES (1793-1868)

Notes on the Western States. Philadelphia; H. Hall

Said erratics and other debris were the result of violent currents washing suddenly submerged land
FOHYAG 628

HAYES, GEORGE EDWARD (1802-82)

Remarks on Geology and Topography of Western New York, in American Journal of Science, 35: 86-105

Said New York was sea floor which had been shaped by waves and currents before being uplifted
GREG 108

HOPKINS, WILLIAM (1793-1866)

Researches in Physical Geology, in Cambridge Philosophical Society Transactions, 6: 1-84

Said moraines were formed by currents which flowed around suddenly uplifted domes
CHOR 226

Said when land uplifted, surface tensions caused longitudinal fracturing followed by transverse cracking at an angle to the fractures
CHR 174-5

LYELL, CHARLES (1797-1875)

Elements of Geology. London; J. Murray

Said corals built their circular reefs atop subsiding volcanoes but horseshoe-shaped reefs were scuplted by a series of earthquakes
STODD 203, 206

Said it was possible for a species to change somewhat in response to its environment
WOOD 107-8

Said metalliferous deposits occurred at the junction of stratified layers and plutonic rocks by sublimation
HTOD 55

Said strata could be dated by studying the order, mineral composition, fragments, and especially the fossils, in each stratum
CONK 80

The Silurian System. London; J. Murray (1839)	Divided England and Wales into eight sedimentary systems based on size, lithology and fossil content CONK 80
	Named his formations after the geographic location where each was best studied for identification and sequence in their series CONK 79
	Said erratics had been encased in ice and floated to their current location where the ice melted and dropped them CGT 13
	Said intrusives, lavas, and ashes had been deposited on ancient seafloors GEI 420
	Said some species vanished after a time, or suddenly when changes destroyed them, while others persisted CONK 80
	Said there was an overall gradual and uniform geologic succession resulting from igneous eruptions which provided the raw materials for deposition CONK 81

SEDGWICK, ADAM (1785-1873)

Seventh Meeting of the British Association for the Advancement of Science, in American Journal of Science, 33: 288	Said valleys and basins had formed originally on the seafloor but had been subsequently uplifted GREG 108

SEFSTROM, NIL GABRIEL (1787-1845)

Untersuchungen über die auf den Felsen Scandinaviens in bestimmter Richtung vorkommenden Furchen und deren wahrschein-bicher Enstheung, in Poggendorff´s Annalen der Physik, 43: 533-70	Said erratics were moved by great ocean currents GIA 159

WHITTLESEY, CHARLES (1808-86)

His Report, in *Second Annual Report of the State of Ohio*: 41-71. Columbus; Samuel Redary

Said drift was stratified
EDEKH 119

1839

BAER, KARL ERNST VON (1792-1876)

Nachricht von der Wanderung eines sehr grossen Granit-Blockes über den Finnischen Meerbusen nach Hochland, in *Impériale Académie de Sciences de St. Pétersbourg Bulletin*, 5: 54-158

Said striae and grooves on surfaces of erratic rocks proved they had been moved by flowing ice
TDG 379

BECHE, HENRY THOMAS DE LA (1796-1855)

Report on the Geology of Cornwall, Devon, and West Somerset. London; Longman, Orme, Brown, Green and Longmans

Said veins formed by lateral secretion
HTOD 57

CONRAD, TIMOTHY ABBOTT (1803-77)

Observations on Characteristic Fossils and upon a Fall of Temperature in Different Geological Epochs, in *American Journal of Science*, 35 (2): 237-51

Said changes in the mean temperature of the earth´s crust had caused extinctions and, subsequently, new species
AIGT 8

Said faunal correlation existed between the Silurian glaciers of Europe and the New York glaciers
SCHU 72

Said glacier ice had moved, polishing and scratching bedrock and transporting erratics
AIGT 9; FAIR 156

Said there had been giant ice sheets
CHOR 270

Said uplift beneath glaciers had sent debris tumbling down the glacier to form moraines
AIGT 8

Said when the temperature had fallen in the past, lakes had frozen and joined to form glaciers
AIGT 9; FOHYAG 203

EHRENBERG, CHRISTIAN GOTTFRIED (1795-1876)

Ueber die Dysodil genannte Mineralspecies als ein Product aus Infusorienschalen, in Poggendorff's Annalen der Physik und Chemie, Series 2, Volume 48, Part 4, Number 12: 573-5 (1839)

Said diatoms were source of bitumen and asphalt
HAN 555; OWEN 32

JACKSON, CHARLES THOMAS (1805-80)

Reports on the Geology of the State of Maine and on the Public Lands Belonging to Maine and Massachusetts, in American Journal of Science, 36: 153

Said that the ancient, mile deep ocean had once flowed inland depositing the erratics
GREG 115

MILLER, WILLIAM HOLLOWES (1801-80)

Treatise on Crystallography. Cambridge; J. and J.J. Deighton

Used spherical trigonometry to describe crystals
DEAS 138

PHILLIPS, JOHN (1800-74)

Treatise on Geology, Volume 1. London; Longman, Orme, Brown, Green and Longman

Suggested that undulations of the crust were caused by 'agitated' liquid beneath the surface
GER 34

1839-40

PREVOST, LOUIS CONSTANT (1787-1856)

Opinion sur la Theorie des Soulèvements, in Société Géologique de France Bulletin, Old Series, 10: 430 and 11: 138-203 (1840)

Said mountains were uplifted by land sinking elsewhere
PSG 138

1840

AGASSIZ, LOUIS (1807-1873)

Etudes sur les Glaciers. Neuchâtel; Jent et Gassmann

Said Europe was populated by elephants and tropical vegetation before the great ice sheet chilled it
FENT 121

Said ice covered the Alpine region before uplift occurred, then rising rock fractured, falling on the ice, and was transported by the process of dilation, which was fastest along the margins
CGT 18; GIA 161; FOHYAG 623-4

On Glaciers, and the Evidence of Their Having Once Existed in Scotland, Ireland, and England, in Geological Society of London Proceedings, 3: 327-332 (1842)

Said large sheets of ice, had covered all of the British Isles
D&B 425; WOOD 152

DARWIN, CHARLES ROBERT (1809-82)

Life and Letters of Charles Darwin, Volume 2. London; J. Murray (1887)

Said earth in perihelion (winter) would react more strongly to the pull of the moon causing floating crust to experience more earthquakes
DAV 57-8

DELAFOSSE, GABRIEL (1796-1878)

Recherches Relatives à la Cristallisation Considérée sous les Rapports Physiques et Mathématiques, in Académie des Sciences, Paris. Comptes Rendus, 11: 394-400

Said crystal shapes were determined by an internal lattice system
WMC 76

LOGAN, WILLIAM EDMOND (1798-1875)

On the Character of the Beds of Clay Lying Immediately Below the Coal Seams of South Wales; and On the Occurrence of Coal-Boulders in the Pennant Grit of That District, in Geological Society of London Transactions, 6: 491-8 (1842)

Said that coal was of vegetable origin and grew in situ
ALCOCK 13

LYELL, CHARLES (1797-1875)

On the Geological Evidence of the Former Existence of Glaciers in Forfarshire, in Geological Society of London Proceedings, 3: 337-45

Said vast glaciers had covered northern Europe leaving erratics, moraines, and sandy deposits behind when they melted
CGT 22

Principles of Geology (Sixth Edition). London; J. Murray

Said new species might appear successively, but not progressively, although accidents of preservation might suggest the latter
SIMP 265; RUD 181, 228

1840

RENDU, LOUIS (1789-1859)

Théorie des Glaciers de la Savoie (Theory of the Glaciers of Savoy), in Académie des Sciences, Belles-Lettres et Arts de Savoie, Chambery, Mémoires, 10: 39-159 (1841)

Said the center of a glacier moved faster than its sides
ROWL 190

ROGERS, HENRY DARWIN (1809-66)

Annual Report on the Geological Survey of Pennsylvania. Boston/Harrisburg; N.P.

Said mountain uplift was the result of both vertical and horizontal forces
GER 31

SOMERVILLE, MARY (1780-1872)

On the Connexion of the Physical Sciences (Fifth Edition). London; J. Murray

Said interior of the earth was a great cavern
NCDAFG 231

1840-41

BUCKLAND, WILLIAM (1784-1856)

Glaciers and the Evidence of Their Having Once Existed in Scotland, Ireland and England, in Geological Society of London Proceedings, 3: 327-32

Said moraines had been deposited by glaciers
CHOR 216, 291

Said erratics had been carried by glaciers
CHOR 216, 291

1841

ABICH, HERMANN WILHELM (1806-86)

Ueber die Natur und Zusammenhang der vulkanischen Bildungen, in Geologische Beobachtungen ueber die vulkanischen Erscheinungen und Bildungen in unter- und mittel-Italien, Volume 1, Part 1. Brunswick; Friedrich Vieweg und Sohn

Proposed igneous rocks be classified in accordance with their feldspar composition
CIR 45; HSP 31

BUCKLAND, WILLIAM (1784-1856)

Anniversary Address of the President, February 19, in Geological Society of London Proceedings, Volume 3, Part 2, Number 72: 332-7, 345-8

Said erratics and drift were deposited by floating ice and glaciers, the latter at higher elevations
EHGT 140; GIA 160

On the Glacio-Diluvial Phenomena in Snowdonia and the Adjacent Parts of North Wales, in _Geological Society of London Proceedings_, 2: 579-88 (1842)

Said moraines and erratics could be cumulative result of glaciers, icebergs, and rivers
CHOR 223

CHARPENTIER, JOHANN VON (JEAN G. F. DE) (1786-1855)

Essai sur les Glaciers et sur le Terrain Erratique du Bassin du Rhône. Lausanne; M. Ducloux

Said during the glacial age, the Alps uplifted, fracturing the strata and admitting water which changed to steam and blocked the sun´s rays, thus causing cold humid conditions, more snow, vast sheets of ice, for an extended period of time, and the extinction of life
CGT 4, 18; GIA 162

DAUBREE, GABRIEL AUGUSTE (1814-96)

Sur le Gisement, la Constitution, et l´Origine des Amas de Minerai d´Etain, in _Annales de Mines_, 29: 65-112

Said fluorine, etc., content of ascending waters determined ore-carrying capability
HTOD 72

DUROCHER, JOSEPH MARIE ELISABETH (1817-60)

Recherches sur les Roches et les Minéraux les Iles Feroé, in _Annales des Mines, Series 3_, 19: 547-92

Said there were five subgroups of igneous rock
HSP 62

HITCHCOCK, EDWARD (1793-1864)

First Anniversary Address before the Association of American Geologists, in _American Journal of Science_, 41 (2): 232-75

Said major valleys and mountain passes were structural in origin and that even subordinate folds or faults could cause minor features
GREG 109

Said overturned strata were the result of horizontal compression
GKES 153

Said the erratics, drift and rounded mountains of New England were result of a southerly moving ice sheet
AIGT 13-4; CHOR 270-5, 294

Sketch of the Scenery of Massachusetts. Northampton; J. H. Butler (1842)

Said erratics and drift had come from the Arctic region on waves of translation (in postscript called it glaciofluvial force)
AIGT 15-6; HLF 159-60

JACKSON, CHARLES THOMAS (1805-1880)

Report on the Geology of the State of New Hampshire, in Silliman Journal (American Journal of Science), 41: 383-4

Said glacial drift came in a sea of ice from the Arctic, prior to the coming of man, during a period of subsidence
HLF 160; MERR 359

Said steam and gases confined in the earth´s molten interior distorted deposits nearby, thus causing inclined strata
MERR 359

Said volcanoes were the molten interior of the earth´s safety valves
MERR 359

MACLAREN, CHARLES (1782-1866)

The Glacial Theory of Professor Agassiz of Neuchâtel, in American Journal of Science, 42 (2): 346-65 (1842)

Said sea level fell eight hundred feet during continental glaciation
GIA 161

MILLER, HUGH (1802-1856)

Old Red Sandstone, or New Walks in an Old Field. Edinburgh; John Johnstone

Said there was evidence of degeneration in some fossil lines, but none of progression
FENT 201

MILNE, DAVID (1805-90)

Notices of Earthquake Shocks Felt in Great Britain, and Especially in Scotland, with Inferences as to the Causes of Such Shocks, in Edinburgh New Philosophical Journal, 31: 262, 275-7

Said vibratory waves travel vertically upward from the epicenter, then outwards in spherical waves, becoming more horizontal as the distance increases
AD 421-2; DAV 43-4

ROUILLIER, KARL FRANTSEVICH (1814-58)

Doubts in Zoology as a Science, in Otechestvennyia Zapiski, Volume 19, Part 2

Said all organisms responded to changes i their environment by changing physically so paleozoogeographical provinces are identifiable
TDG 373

STEENSTRUP, JOHANNES IAPETUS SMITH (1813-97)

Geognostisk-Geologisk Undersögelse af Skovmoserne Vidnesdam og Lillemose i det Nordlige Sjelland, in Kongelige Danske Videnskabernes Selskab Skrifter (1842)

Said fossil pollen could be used to distinguish stratigraphical succession
HCMP 282

1841

SURELL, ALEXANDRE

Etude sur les Torrents des Hautes Alpes. Paris; L'Auteur

Said streams develop a longitudinal profile, which is (geometrically) a critical curve composed of successive slopes designed for maximum discharge
CHOR 283-7

Said streams erode upstream and deposit the debris downstream in an attempt to achieve grade
CHOR 283-7

Said rivers pass through three stages: 1) erosion and deposition, 2) meandering, and 3) grade
CHOR 283-7

Said soil conservation was a serious problem
MATH 372

1842

AGASSIZ, LOUIS (1807-73)

Théorie des Glaces et ses Progrès les Plus Récents, in Edinburgh New Philosophical Journal, 33: 217-83

Said that the center of a glacier moved faster than the sides
ROWL 190

FRANKENHEIM, MORITZ LUDWIG (1801-69)

System der Krystalle. Breslau; Druck von Grass, Barth

Described the internal arrangement of crystals by use of fourteen space lattices
PHIL 225-6

HITCHCOCK, EDWARD (1793-1864)

Phenomena of Drift, or Glacio-aqueous Action in North America, between the Tertiary and Alluvial Periods, in Association of American Geologists and Naturalists Reports, 1842: 164-221

Said North American topography was result of a glacier plus subsequent water action
FOHYAG 625-6

HOPKINS, WILLIAM (1793-1866)

On the Elevation and Denudation of the District of the Lakes of Cumberland and Westmorland, in Geological Society of London Proceedings, 3: 757-66

Said the sudden uplifting of land threw the overlying sea back causing tidal waves which carried rocks and soil, and deposited erratics and other drift
CHOR 225-6, 340

On the Thickness and Constitution of the Earth's Crust, in *Royal Society of London. Philosophical Transactions*, 132: 43-56

Said earth had solid shell at least 800 or 1000 miles thick and the core or entire earth might be solid
BRUSH 707; NCDAFG 232-3

Said lake of fluid inside the earth fed volcanoes
NCDAFG 233

MURCHISON, RODERICK IMPEY (1792-1871)

Anniversary Address of the President, February 18, in *Geological Society of London Proceedings*, 3: 637-87

Said that striae were a structural phenomenon
CHOR 228

Said glaciers had been active only high in mountains in northern lands and lowland erratics were the result of icebergs, continental meltwater, and oceanic currents
EHGT 140; GIA 162-3

Said most gravel, boulders, and clay were deposited in the ocean in the past
CHOR 227-8, 295

MURCHISON, RODERICK IMPEY AND ALEXANDER KEYSERLING AND EDOUARD VERNEUIL (1792-1871 / 1815-91 / 1805-73)

Memoir on the Geological Structure of the Ural Mountains, in *Geological Society of London Proceedings*, 3: 742-53 (1838-43)

Said mineral veins were associated with eruptive rock
HTOD 57

ORBIGNY, ALCIDE DE (D'ORBIGNY) (1802-57)

Terrains Jurassique, Paléontologie Française. Paris; Masson

Divided strata into twenty-seven units, or stages, based on fossil changes
CONK 82

PERCIVAL, JAMES GATES (1795-1856)

Report on the Geology of the State of Connecticut. New Haven; Osborn and Baldwin

Said peneplains were an effect of structur
GREG 111

ROGERS, HENRY DARWIN AND WILLIAM B. ROGERS (1808-66 / 1804-82)

On the Physical Structure of the Appalachian Chain, as Exemplifying the Laws Which Have Regulated the Elevation of Great Mountain Chains Generally, in Reports of the First, Second, and Third Meetings of the Association of American Geologists and Naturalists at Philadelphia in 1840, and at Boston in 1842: 474-531

Said glacial drift and striations were caused by wave-like motions of the ocean floor which propelled water and detritus onto the land
GER 29

Said mountains were uplifted when gas and vapor induced cracking of the crust and repeated waves in the molten liquid below exerted vertical and horizontal forces
JORD 98; GER 26-8

On the Physical Structure of the Appalachian Chain, as Exemplifying the Laws Which Have Regulated the Elevation of Great Mountain Chains Generally, in British Association for the Advancement of Science Report Part 2: 40-2

Said Appalachia was uplifted by expansion of magma and gases
AD 396-7

On the Physical Structure of the Appalachian Chain, as Exemplifying the Laws Which Have Regulated the Elevation of Great Mountain Chains Generally, in American Journal of Science, 43: 177-9

Said folded strata resulted from elastic vapors rising through parallel cracks in the earth and causing waves in the fluid beneath the crust
GKES 154

ROGERS, HENRY DARWIN (1809-66)

An Inquiry into the Origin of Appalachian Coal Strata, Bituminous and Anthracitic, in Reports of the First, Second, and Third Meetings of the Association of American Geologists and Naturalists at Philadelphia in 1840, and at Boston in 1842: 474-531

Said coal was volatized by, and in proportion to, the heat released by subterranean vapors during structural folding
GER 31

Said coal formed in continental borderlands which were alternately drained during uplift, drowned during subsidence, and marshy other times
GER 30-1

VANUXEM, LARDNER (1792-1848)

Geology of the Third New York District. Albany; Carroll and Cook

Said drift was transported and deposited by local ice
HLF 160

1842-43

EMMONS, EBENEZER (1799-1863)

Geology of New York. Albany; Carroll and Cook

Said rock scorings and fine drift had been caused by shallow river currents but coarser drift and erratics had been transported by ice
HLF 160

1843

BURAT, AMEDEE (1809-83)

Géologie Appliquée (Applied Geology). Paris; Langlois et Leclerc

Said copper, lead and iron were deposited by eruptive action
TODH 322

DANA, JAMES DWIGHT (1813-95)

On the Analogies between the Modern Igneous Rocks and the So-called Primary Formations, and the Metamorphic Changes Produced by the Heat in the Associated Sedimentary Deposits, in Silliman Journal (American Journal of Science), 45: 104-29

Said most granites were of igneous origin
GKES 143

Said sedimentary formation did not cause schistocity
GKES 143

Said volcanic fires heating seawater might account for schistose structure of gneiss and mica-slate, and might also metamorphose granites
FOHYAG 240-1

MATHER, WILLIAM WILLIAMS (1804-1859)

Geology of New York. Albany; Carroll and Cook

Said sudden collapse of earth's crust caused great currents which carried ice and debris which was subsequently dropped, thus accounting for drift phenomena
HLF 160

ROGERS, WILLIAM B. AND HENRY DARWIN ROGERS (1804-82 / 1808-66)

On the Physical Structure of the Appalachian Chain, in Association of American Geologists and Naturalists Transactions: 474-53

Said folded mountains were uplifted by tangential forces and upward billowing movements
PSG 138

1844

CHAMBERS, ROBERT (1802-1871)

Vestiges of the Natural History of Creation. London; J. Churchill

Said new species were created by the demands of the environment and by trans-specific jumps
RUD 205

Said higher forms of life developed naturally from lower forms
FENT 201-2

DARWIN, CHARLES ROBERT (1809-82)

Foundations of the Origin of Species. Cambridge; University Press (1909)

Said life evolved through process of natural selection
RUD 232

Geological Observations on the Volcanic Islands and Parts of South America Visited During the Voyages of the HMS Beagle, with Brief Notices on the Geology of Australia and the Cape of Good Hope, Being the Second Part of the Voyage of the "Beagle", During 1832-36. London; Smith, Elder

Said that some valleys and water gaps had been cut by waves, tides or other sudden currents
GREG 109

Said igneous rock was either 1) volcanic or 2) plutonic in origin, and 1) acid or 2) basic in composition
HSP 17

DUFRENOY, OURS PIERRE ARMAND PETIT (1792-1857)

Traité de Minéralogie. Paris; Carilion-Goeury et Vor Dalmont (1844-47)

Said the only basis for identifying a mineral species was chemical composition
WMC 72

HUMBOLDT, FRIEDERICH HEINRICH ALEXANDER VON (1769-1859)

Kosmos (Cosmos). Stuttgart und Tubingen; J. G. Cotta (1845-58) and London; H. G. Bonn (1848-65)

Said that petroleum was distilled by volcanic action from deeply buried strata
OWEN 30

Said outward rippling earthquake waves may intersect and cancel each other out
AD 422

Said earthquakes were waves of vibration
DAV 68

ROGERS, HENRY DARWIN (1808-1866)

Brief History of the Recent Labors of American Geologists and a Rapid Survey of the Present Condition of Geological Research in the United States, An Address Delivered at the Meeting of the Association of American Geologists and Naturalists, Held in Washington, May, 1844, in _American Journal of Science_, 47: 137-60, 247-78

Said earthquake-propelled waves of northern or circumpolar waters, moved southward by polar uplift and subsidence to the south, deposited the erratics
GREG 118; HLF 161

A Communication, in _Abstract of the Proceedings of the Fifth Session of the Association of American Geologists and Naturalists_. New York; Wiley and Putnam

Said there was six times as much carbon taken from the air and stored in coal deposits as there is in the present atmosphere
FAIR 42

STUDER, BERNARD (1794-1887)

Lehrbuch der physikalischen Geographie und Geologie. Bern; J. F. J. Dalp

Said magma forced its way surfaceward, upturning strata above and folding strata aside to form an igneous core surrunded by crumbled sedimentary strata
BEL 34-5

1844-45

FOURNET, JOSEPH JEAN BAPTISTE XAVIER (1801-69)

Essai sur les Filons Métallifères du Département de l´Aveyron, in _Annales des Sciences Physiques et Naturelles, d´Agriculture_, 7: 1-88

Said there were three types of intrusive ore veins, each identifiable by its particular igneous origin
HTOD 51

1845

BURAT, AMEDEE (1809-1883)

Théorie des Gîtes Métallifères. Paris; Langlois et Leclerc

Said water and vapors rising from igneous intrusives were necessary for formation of ore deposits
HTOD 51

)RBES, JAMES DAVID (1810?-)

Papers on Glaciers, in Royal Society of Edinburgh Proceedings, 1: 406-7, 409-11, 414-15

Said glaciers moved as if they were composed of viscous liquids
ROWL 190

:TCHCOCK, EDWARD (1793-1864)

Elementary Geology (Third Edition). New York; M. H. Newman

Said bitumins were produced from vegetable matter, the same way coal was, and were driven surfaceward by the earth´s internal heat
OWEN 41-2

Said drift was result of uplift, earthquake vibrations, glaciers, and iceberg transport
AIGT 19; GIA 165

ATHER, WILLIAM WILLIAMS (1804-1859)

On the Physical Geology of the United States East of the Rocky Mountains, in American Journal of Science, 49 (1): 1-20, 284-301

Said uplift was caused by contraction of strata due to cold
FOHYAG 252-3

Said plant debris, carried by currents from the tropics, had been trapped in eddies, deposited, and turned to coal
FOHYAG 252-3

Said glacial detritus had been water-borne
FOHYAG 252-3

ERREY, ALEXIS (1807-82)

Mémoire sur les Tremblements de Terre dans le Bassin du Rhin, in Annales des Sciences Physiques et Naturelles, d´Agriculture, 8: 265-346

Said the epicenter of an earthquake was an irregular, jerking, tumultuous area where there was no prevailing wave direction discernible
DAV 56

845-46

JROCHER, JOSEPH MARIE ELISABETH (1817-1860)

Etude sur le Métamorphisime des Roches, in Société Géologique de France Bulletin, 3: 546-647

Said there was a difference between contact and regional metamorphism
HSP 57

1846

DANA, JAMES DWIGHT (1813-95)

On the Early Conditions of the Earth's Surface, in Silliman Journal (American Journal of Science), 2: 347-8

Said as earth cooled, areas without volcanoes solidified first, then, increasing cooling caused contractions forming ocean basins and revealing ancient mountains and continents
FOHYAG 259

Volcanoes of the Moon, in American Journal of Science, Second Series, 2: 335-55

Said the earth's cooling crust had solidified unevenly and subsequent subsidence of the ocean basins caused lateral pressures which folded thick deposits and raised mountains
DRH 242

DARWIN, CHARLES ROBERT (1809-82)

Geological Observations in South America. London; Smith, Elder and Co.

Said the direction of veins depended on the structure of the rocks, the degree o metamorphism and the presence of intrusi
HTOD 57-8

FORBES, JAMES DAVID (1810?-)

Part I. Containing Experiments on the Flow of Plastic Bodies, and on the Phenomena of Lava Streams, in Royal Society of London Philosophical Proceedings, 136: 143-56

Said that the center of a glacier moved faster than the sides
ROWL 190

MALLET, ROBERT (1810-81)

On the Dynamics of Earthquakes, in Royal Irish Academy Transactions, 21: 50-106 (1848)

Said an earthquake was a wave of elastic compression transmitted omnidirectionall from one or more centers
GEI 277-8

Said earthquakes were waves of vibration which could be affected by uplift or subsidence
DAV 68

Said the velocity of earthquake waves was affected by the elastic constant of rocks, thus providing a key to the identification of substrata
DAV 69

AMSAY, ANDREW CROMBIE (1814-91)

Denudation of South Wales, in Memoir of the Geological Survey of Great Britain, 1: 297-335

Reconstructed original stratification from folded, unconformable structures
CHOR 304, 463

Said rivers had little erosive power
CHOR 304

Said subsiding and inundated land masses were leveled by the sea and the debris eventually formed new strata
CHOR 304, 463

Said that peneplains were the result of marine erosion
GREG 111

CHEERER, THEODOR (1813-75)

Discussion sur la Nature Plutonique du Granite et des Silicates Crystallins Qui s´y Rallient, in Société Géologique de France Bulletin, 4 (1): 468-95 (1846-47)

Said granite magmas were result of aqueous solutions exuding from granite intrusions
HTOD 58

846-47

ISCHOF, KARL GUSTAV (1792-1870)

Lehrbuch der chemischen und physikalischen Geologie (Textbook of Chemical and Physical Geology). Bonn; Marcus

Said salt deposits were result of evaporation in restricted seas
OWEN 34

847

EAUMONT, LEONCE ELIE DE (1798-1874)

Note sur les Emanations Volcaniques et Métallifères (Volcanic and Metalliferous Emanations), in Société Géologique de France Bulletin, Volume 4, Part 2: 1249-1333

Said metalliferous veins and igneous dikes constituted a petrological group
HTOD 60

Said minerals were deposited in veins and mineral waters by vapors from underground water heated by eruptives and the earth´s internal heat
HTOD 60; TODH 322-3

Said there are two kinds of volcanic products: lavas and sublimates and in all volcanic activity there is much water vapor and volatiles
HSP 73; HTOD 60

1847

Said volcanic vapors bearing ore materia deposit them in different rocks dependin on the chemical properties of the metals involved
POGS 124; HTOD 60

DANA, JAMES DWIGHT (1813-95)

Origin of the Grand Outline Features of the Earth, in *Silliman Journal (American Journal of Science)*, 3: 381-99

Said relief features were formed by the contracting of the crust when the earth solidified
FAIR 55

Geological Results of the Earth´s Contraction in Consequence of Cooling, in *American Journal of Science*, 3: 176-88

Said topography was result of lateral pressure overturning crust in its path as it was forced away by consolidation during cooling
FAIR 55; GKES 155

DELESSE, ACHILLE ERNEST OSCAR JOSEPH (1817-81)

Procèdé Mécanique pour Déterminer la Composition des Roches, in *Académie des Sciences, Paris. Comptes Rendus*, 25: 544-5

Said geometrical analyses of the mineral composition of rocks could be determined by areal measurement
HSP 30

HOPKINS, WILLIAM (1793-1866)

Report on the Geological Theories of Elevation and Earthquakes, in *British Association for the Advancement of Science Report* 33-92

Said focus of an earthquake would be the right angle intersection of two lines of direction, at the surface, at the moment motion began
DAV 83

NOGGERATH, JOHANN JAKOB (1788-1877)

Das Erdbeben vom 29 Juli 1846 in Rheingebiet und den Benachbarten. Bonn; Henry und Cohen

Located an epicenter by drawing isoseism lines
DAV 122

PERREY, ALEXIS (1807-82)

Memoir to Royal Academy of Belgium, in *Society Memoirs*, 36: 537 (1853), 52: 146+ (1861), and 81: 690+ (1875)

Said variations in the number of earthqu was due to 1) relative directions from earth of the sun and moon, 2) the distanc from earth to the moon, and 3) the crossi of the meridian by the moon
DAV 58

1847

HEWELL, WILLIAM (1794-1866)

On the Wave of Translation in Connection with the Northern Drift, in Geological Society of London Quarterly Journal, 3: 227-32

Said the European drift deposits were result of a succession of waves of translation (paroxysms)
CHOR 340-1; GREG 119

847-54

ISCHOF, KARL GUSTAV (1792-1870)

Lehrbuch der chemischen und physikalischen Geologie (Textbook of Chemical and Physical Geology). London; Cavendish Society (1854-59)

Said low-temperature aqueous processes were important in formation of rocks and ore deposits, the latter being formed by lateral secretion
HTOD 67-9

848

WEN, RICHARD (1804-92)

On the Archetype and Homologies of the Vertebrate Skeleton. London; R. and J. E. Taylor

Said all bones of mammals were identifiably similar to those in the other vertebrate classes since all derived from a single ideal archetype
RUD 211

848-49

HAMBERS, ROBERT (1802-1871)

Ancient Sea Margins as Memorials of Changes in the Relative Level of Sea and Land. Edinburgh; W. and R. Chambers

Said terraces resulted, 1) when sea level fell leaving old beaches high and dry, or 2) when a sediment-filled lake began cutting back through its bed
CHOR 315-8

849

ANA, JAMES DWIGHT (1813-95)

Narrative of the United States Exploring Expedition During the Years 1838-1842 under the Command of Charles Wilkes; Volume 10 (Geology of the Pacific A). Philadelphia; C. Sherman

Said crystallization of igneous rocks was preceded by a period of liquation-differentiation
HSP 72

Said coral reefs were result of a reef encircled island subsiding
FENT 212; MERR 423

Said the cooling earth is still contract
and the resultant stresses are raising
mountains
MERR 426

Said temperature influenced coral growth
and occurrence
MERR 423

Review of Chamber's 'Ancient Sea Margins' with Observations on the Study of Terraces, in American Journal of Science, Second Series, 7: 1-14, 8: 86-9

Said most valley erosion is the result o
rain cumulating into strong streams
CHOR 363-4

Said topography of Pacific islands was
result of 1) volcanic, or other internal
action, 2) erosion by the sea, 3) weathe
by rain, and 4) destruction by vegetatio
and the elements
CHOR 363-4

Said terraces were result of sea level
falling thus lowering the rivers also
CHOR 318

HITCHCOCK, EDWARD (1793-1864)

On the River Terraces of the Connecticut Valley and On the Erosions of the Earth's Surface, in American Association for the Advancement of Science Proceedings, 2: 148-56 (1850)

Said there were three kinds of terraces
identifiable by their origins: 1) relics
of ancient sea-margins, or raised beache
2) margins of fresh-water lakes; or
3) margins of rivers
CHOR 315

NAUMANN, KARL FRIEDRICH (1797-1873)

Lehrbuch der Geognosie. Leipzig; W. Engelmann

Classified rocks as 1) crystalline,
2) clastic, or 3) other
CROSS 356-7; HTOD 158-9

PETIT, F.

Sur la Densité Moyenne de la Châine des Pyrénées, in Académie des Sciences, Paris. Comptes Rendus, 29: 729-34

Said voids inside of mountains accounted
for their mass being less than had been
expected
DALY 2

QUENSTEDT, FRIEDRICH AUGUST VON (1809-89)

Atlas zu Cephalopoden. Tubingen; L. F. Fues

Said that the origin of oil was from
decaying animal material
OWEN 33

1849

ORBY, HENRY CLIFTON (1826-1908)

On the Microscopical Structure of the Calcareous Grit of the Yorkshire Coast, in West Yorks Geological Society of London Proceedings, 3: 197-205

Said minerals in rocks could be identified by slicing the rock and examining it under the polarizing microscope
GRA 37; JUDD 196

HOMSON, JAMES (1822-92)

Theoretical Considerations on the Effect of Pressure in Lowering the Freezing Point of Water, in Royal Society of Edinburgh Transactions, 16: 575-80

Said changes in equal hydrostatic pressure have direct and mathematically predictable effects on the freezing point of water
ETHP 369

SIGLIO, J.

Analyse de l´Eau de la Méditerranée sur les Côtes de France, in Annales de Chemie, 27: 92-107, 172-91

Said sea water contained minerals which precipitated in a set sequence
P&S 89

1850

ADAMS, CHARLES BAKER (1808-60)

Suggestions on Changes of Level in North America During the Drift Period, in American Association for the Advancement of Science Proceedings: 60-64

Attributed southward movement of glaciers to northern land uplift
HLF 163

AGASSIZ, LOUIS (1807-73)

Lake Superior; Its Physical Character, Vegetation and Animals, Compared with Those of Other and Similar Regions. Boston; Gould, Kendall and Lincoln

Said that northeastern America and northwestern Europe were glaciated by the same huge ice sheet
AIGT 23

BRAVAIS, AUGUSTE (1811-63)

Mémoire sur les Systèmes Formés par des Points Distribués, Régulièrement sur un Plan ou dans l´Espace, in Journal of the Ecole Polytechnique, Cahier 33, 19: 1-128

Said the crystal forms which tend to occur most frequently are those with faces parallel to planes with the smallest reticular area
PHIL 80

CLAUSIUS, RUDOLF (1822-88)

Notiz über den Einfluss des Drucks auf das Gefrieren der Flüssigkeiten, in Annalen der Physik, 81(2): 168-72	Said that if a high-temperature phase had greater volume, when the latent temperature of the phase change exceeded zero degrees, then pressure would increase the temperature NCDAFG 237-8

DANA, JAMES DWIGHT (1813-95)

On Denudation in the Pacific and On the Degradation of the Rocks of South Wales and Formation of Valleys, in American Journal of Science, 9: 48-62 and 9: 289-94	Said that valleys had been cut by streams and that many bays, inlets and fiords were drowned mouths of such valleys GREG 109-10
	Said slope and quantity of water determined whether, and where, valley would be 1) narrow gorge with cascades and 30-60 degree walls, 2) narrow valley with vertical walls and meandering stream, or 3) wide with extensive plain CHOR 365-6
System of Mineralogy (Third Edition) New York; George P. Putman	Classified minerals according to their chemical characteristics GRA 35-7; RPM 12

HUBBARD, OLIVER PAYSON (1809-1900)

On the Condition of Trapp Dikes in New Hampshire, an Evidence and Measure of Erosion, in American Journal of Science, 9: 158-71	Said the valleys in New Hampshire had been cut by streams GREG 110

LYELL, CHARLES (1797-1875)

Anniversary Addresses of the President, in Geological Society of London Quarterly Journal, 6: 27-66	Said fissures rising from extinct mineral springs to surface springs carried mineral-rich vapors upward, creating metalliferous veins HTOD 55-6
	Said species changed slowly through geological time and abrupt changes between formations were accidental RUD 228

ORBIGNY, ALCIDE DE (D´ORBIGNY) (1802-57)

Recherches Zoologiques, in Annales des Sciences Naturelles, 13: 218-36

Said there had been twenty-seven extinctions followed by creation of new species, though some modification in morphology might occur within a period
CBS 220; D&B 24

THOMSON, WILLIAM (LORD KELVIN) (1824-1907)

Effect of Pressure in Lowering the Freezing Point of Water, Experimentally Demonstrated, in Philosophical Magazine, 37(3): 123-7

Said a high-temperature phase of greater volume, when the latent temperature of the phase change exceeded zero degrees, would increase in temperature with pressure
NCDAFG 237-8

1851

BEARDMORE, NATHANIEL (1816-1872)

Manual of Hydrology
London; Waterlow (1862)

Said stream discharge depended on the amount of rainfall and the type of bedrock involved
CHOR 428-30

Said streams partially regulated their flow by changing shape
CHOR 428-30

Said the speed of a stream was greater at surfaces than near the beds, and in the middle than near banks
CHOR 428-30

BUNSEN, ROBERT WILHELM EBERHARD (1811-1899)

Ueber die Prozesse der vulkanischen gesteins-bildungen Islands, in Poggendorff´s Annalen der Physik, 83: 197-272

Said minerals crystallized out of a magma in an order independent of their order of infusibility
YODER 9

Said two magmas produced all igneous rocks and were 1) trachytic (basic) or 2) pyroxenic (acidic)
CROSS 355; HSP 61

LYELL, CHARLES (1797-1875)

A Manual of Elementary Geology
(Third Edition) London;
J. Murray

Said rocks should be classified according to their age
OLD RHG 244

ORBIGNY, ALCIDE DE (D'ORBIGNY) (1802-57)

Cours Elémentaire de Paléontologie et de Géologie Stratigraphiques Paris; V. Masson (1849-52)

Said there had been five geologic periods divided into 27 stages and estimated the total number of species extant in each to be 18,286
CBS 222

TAYLOR, THOMAS JOHN

An Inquiry into the Operation of Running Streams and Tidal Waters. London; Longman, Brown, Green and Longmans

Said steepness of stream beds would be affected by number and size of particles in the discharge water
CHOR 431

Said stream bed erosion would be proportionate to the rate at which the source area eroded
CHOR 432

Said stream beds which could not deepen would broaden and might form shoals and islands
CHOR 433

1852

BEUDANT, FRANCOIS SULPICE (1787-1850)

Traité Elémentaire de Minéralogie. Leipzig; Engelmann

Said a mineral species was those minerals with the same form, colors, refraction, weight and chemical composition
WMC 71

DESOR, EDOUARD (1811-82)

On the Drift of Lake Superior and Post-Pliocene of the Southern States, in *American Journal of Science*, 13: 93-109 and 14: 49-59

Said boulders on Long Island had been transported there by ocean ice
GREG 109

EMMONS, EBENEZER (1799-1863)

Report of Professor Emmons on His Geological Survey of North Carolina. Raleigh; S. Gales

Classified rocks as 1) pyrocrystalline, 2) pyroplastic, or 3) hydroplastic
MERR 431

HOPKINS, WILLIAM (1793-1866)

Annual Address, in *Geological Society of London Quarterly Journal*, 8: 21-80

Said drift had been deposited out of currents generated by repeated uplifts
GREG 119

Said erratics might be result of glaciers, floating ice, or currents
CHOR 331

Said slow cooling of the earth dictated future geological changes
NCDAFG 231

LYELL, CHARLES (1797-1875)

A Manual of Elementary Geology (Fourth Edition) London; J. Murray

Said intermittent uplift and river cutting produced terraces
CHOR 184-5, 294

ROSE, GUSTAV (1798-1873)

Das krystallo-chemische Mineralsystem. Leipzig; Engelmann

Said mineral species were determined by the elements involved and their final chemical composition
RMP 12; WMC 72

1853

GREENWOOD, GEORGE (1799-1875)

Tree-Lifter. London; Longman, Brown, Green and Longmans

Said all erosion could be explained by rain and rivers
CHOR 373

PHILLIPS, JOHN (1800-74)

The Rivers, Mountains, and Sea-Coast of Yorkshire London; J. Murray

Said the atmosphere, carbonic acid, and precipitation weather the land and the rivers, then carry the debris away through sea-cut valleys
CHOR 320

Said topography was determined by rock composition and structure
CHOR 320

Said uplifted land broke apart, fell into the sea, was weathered, and then uplifted again
CHOR 318-20

THURMANN, JULES (1804-55)

Résumé des Lois Orographiques Générales du Système des Monts-Jura, in Actes de la Société Helvétique des Sciences Naturelles, 38th Session, Porrentruy: 280-92

Said upward warping of strata was result of lateral compression of sediment cover
DGGS 50

1853

WALTERSHAUSEN, WOLFGANG SARTORIUS VON (1809-76)

Ueber die vulkanischen Gesteine in Sicilien und Island und ihre submarine Umbildung
Gottingen; Dieterichshen Buchhandlung

Said magma was feldspar-rich above and graduated to pyroxine or magnetite-rich below, depending on the specific gravity of the minerals involved
HSP 62

1854

EMMONS, EBENEZER (1799-1863)

American Geology. Albany; Sprague and Co.

Said earth´s heat raised areas by causing internal expansion, transferred fused material to the surface, and caused subsidence by destroying strata below
FOHYAG 335

Said the age of a rock was the age at which it crystallized, the oldest having been hottest and therefore more highly fused and crystalline
FOHYAG 335

HIND, HENRY YOULE (1823-1908)

The Origin of the Basin of the Great Lakes (never published)

Said that glaciers had dug out the Great Lakes
GREG 125

MURCHISON, RODERICK IMPEY (1792-1871)

Siluria. London; J. Murray

Said some bituminous shales were algal in origin
OWEN 32

PRATT, JOHN HENRY (1811-71)

On the Attraction of the Himalaya Mountains and the Elevated Regions Beyond Them Upon the Plumb-line in India, in Royal Society of London, Philosophical Transactions, Series B, 145: 53-100 and in American Philosophical Magazine, 9: 230-35 and 10: 240 (1855)

Said when mountains were raised by expansion, the density of underlying rocks decreased
MAB 61

Said accumulating sediments caused the continents to sink
CGPS 276

Said that the Himalayas caused the density of neighboring rock masses to vary as much as three-fold
DALY 2

WHITNEY, JOSIAH DWIGHT (1819-96)

Metallic Wealth of the United States. Philadelphia; Lippincott, Grambo and Co.

Said decomposition of ores which had formed from sulphurets of iron and copper with a quartzose gangue would yield gossan and cupriferous ore
POGS 127-8

Said when aqueous solutions, concentrated at the surface, sank through strata picking up ore particles, then ore deposits were formed in 1) vertical fissures, unstratified, 2) flat sheets, stratified, or 3) residuals from decomposing limestone
FOHYAG 329; TODH 328

1855

AIRY, GEORGE BEDELL (1801-92)

On the Computation of the Effect of the Attraction of Mountain Masses as Disturbing the Apparent Astronomical Latitude of Stations in Geodetic Surveys, in Royal Society of London, Philosophical Transactions, 145: 101-104

Said earth´s crust floats on lava like a raft of timbers on water and logs which float highest lie deeper in water than others so mountains must have deep roots of light material
AD 397; WOOD-2 252

Said mountain ranges, continents, and large table-lands float on roots of lighter material than do ocean floors but that the composition of mountains also provided self-support
DALY 2

Said earth´s crust floated on lava and when crust thickened, mountains were uplifted
MAB 61

Explained Pratt´s Himalayan problem by isostasy
BRUSH 707; MATH-1 401

DANA, JAMES DWIGHT (1813-95)

On American Geological History, in American Journal of Science, 2: 305-34 (1856)

Said glacial period in America consisted of 1) northern uplift, 2) depression of the land, and 3) moderate reelevation of the land
HLF 163, 168-9

Said glaciers were result of falling temperatures and rising land
HLF 168

KJERULF, THEODOR (1825-88)

Christiania-Silurbecken, chemische-geognostische Untersucht. Christiania; Kongelige Norske Frederiks Universitet

Said batholiths resulted from the gradual assimilation of rock surrounded by an accumulation of magma
HSP 18

PHILLIPS, JOHN (1800-74)

Manual of Geology. London; R. Griffin and Co.

Said streams could cut small channels
CHOR 321

Said transported debris often fills in the valleys of rivers
CHOR 321

RAMSAY, ANDREW CROMBIE (1814-91)

On the Occurrence of Angular, Subangular, Polished, and Striated Fragments and Boulders in the Permian Breccia of Shropshire, Worcestershire, Etc., in Geological Society of London Quarterly Journal, 2: 185-205

Said some drift had been carried by ice
CHOR 334-5

WALLACE, ALFRED RUSSEL (1823-1913)

On the Law Which Has Regulated the Introduction of New Species, in Annals and Magazine of Natural History, Series 2, 16: 184-96

Said that species come into existence coincident in space and time with pre-existing, closely allied species
RUD 226

1856

ABICH, HERMANN WILHELM (1806-86)

Vergleichende chemische Untersuchungen der Wasser des kapischen Meeres, Urmia und Van-see´s, in St. Petersburg Académie Imperiale des Sciences Mémoires, Series 6, Part 1, 7(9): 1-57

Said that analysis of fossil shells would determine the composition of the waters in which they had lived
TDG 380-1

DANA, JAMES DWIGHT (1813-95)

On American Geological History, in American Journal of Science, Second Series, 22: 305-34

Said contractions of the cooling earth forces the oceanic crust to press against the continents resulting in uplift, volcanoes and metamorphism
DRH 252

Said that the earth´s major structural features and the shape of the continents were defined by the natural cleavage lines, i.e., the ocean basins
DRH 252

DARCY, HENRI PHILBERT GASPARD (1803-58)

Les Fontaines Publiques de la Ville de Dijon (The Public Fountains of the City of Dijon) Paris; V. Dalmont

Said the flow rate of groundwater is directly proportional to 1) the energy loss and 2) the type of sand involved and is inversely proportional to the distance traveled
CHOW 8

FERREL, WILLIAM (1817-91)

Essay on the Winds and Currents of the Ocean, in Nashville Journal of Medicine and Surgery (October) and in Professional Papers of the U.S. Signal Service, Number 12, Washington, D.C., 1882

Said that the circulation patterns of the oceans and atmosphere were a result of the earth´s rotation
SOC 149

FOURNET, JOSEPH JEAN BAPTISTE XAVIER (1801-69)

Aperçus Rélatifs à la Théorie des Gîtes Métallifères (and) Aperçus Rélatifs à la Théorie des Filons, in Academie des Sciences, Paris. Comptes Rendus, 42: 1097-1105 and 43: 345-52, 842-49, 894-900

Said ore deposits and alteration were result of atmospheric action on intrusions of magma
HTOD 53

HITCHCOCK, EDWARD (1793-1864)

Description of a Large Boulder in the Drift of Amherst, Massachusetts with Parallel Striae upon Four Sides, in American Journal of Science, Series 2, 27: 397-400

Said alluvial deposits were still forming and were either 1) drift unmodified (results of glaciers, icebergs, landslips and earthquakes) or 2) drift modified (beaches, submarine ridges, sea floors, dunes, terraces, deltas, moraines, eskers)
CHOR 315

LESLEY, JOSEPH PETER (1819-1903)

Manual of Coal and Its Topography. Philadelphia; J.B. Lippincott and Co.

Said that structure determined the shape of all sizes of surface features and that folded strata impress their shape on the surface layers
CHOR 346-54

OPPEL, CARL ALBERT VON (1831-65)

Die Juraformation Englands, Frankreichs und des Sudwestlichen Deutschlands, in Wurttembergische Naturforschende Jahreshefte, 12-14
Stuttgart; Ebner and Seubert

Said to obtain an ideal profile of a stratigraphic column, only the fossil zones should be considered
CBS 221; COX 218

Correlated strata of England, France, Switzerland and South Germany based on their fossils
WSSB 223

PERREY, ALEXIS (1807-82)

Mémoire sur les Tremblements de Terre, in Mémoires Couronnés et Autres Mémoires. Brussels; Académie Royale des Sciences des Lettres et des Beaux Arts de Belgique

Said earthquake tremors were felt simultaneously in several and distant countries
DAV 47-9

ROGERS, WILLIAM B. AND HENRY DARWIN ROGERS (1804-82 / 1808-66)

On the Laws of Structure of the More Disturbed Zones of the Earth´s Crust, in Royal Society of Edinburgh Transactions, 21: 431-72

Said movements of the land were caused by pulsating molten lava beneath the crust of the earth
AD 396-7; GER 28

SCHUROVSKY, G.E.

Oscillating motion of the European Continent during Historical and Nearly-Historical Times. Moscow; N.P.

Said the earth´s interior energy caused volcanic activity, slow uplift, and subsidences
TDG 379-80

VOLGER, GEORG HEINRICH OTTO (1822-97)

Untersuchengen über das Letztjahrige Erdbeben in Central-Europe, in Petermanns Geographische Mittheilungen: 85-102

Mapped earthquake intensities with isoseismal lines corresponding to scale of intensity: 1) buildings fell down, 2) strongest parts of building were damaged, 3) buildings were slightly damaged, 4) little building damage, and 5) shock was felt
DAV 123

357

RONN, HEINRICH GEORG (1800-62)

Natura Doceri (To Be Taught by Nature), in Académie des Sciences, Paris. Comptes Rendus, 44: 166-7

Said extinct species were replaced by higher forms due to natural production of successive species in response to environmental changes
CBS 221; RUD 22-3

ANA, JAMES DWIGHT (1813-95)

Thoughts on Species, in American Journal of Science, 24: 305-16

Said species were permanent but had infinite diversity while hybrids were temporary changes
SCHU 92-3

UROCHER, JOSEPH MARIE ELISABETH (1817-1860)

Essai de Pétrologie Comparée, in Annales des Mines, 11: 217-259, 676-81

Said there were three types of igneous rocks: 1) acid, 2) basic, and 3) hybrid
CROSS 355

Recherches sur les Roches Ignées, sur les Phénomènes de leur Emission et sur leur Classification, in Académie des Sciences, Paris. Comptes Rendus, 44: 325-330+

Said different magmas, one basic and the other acidic, lay at different depths and rock produced depended on whether it was a result of liquation or was a mixture of magmas
HSP 62

REENWOOD, GEORGE (1799-1875)

Rain and Rivers; or Hutton and Playfair against Lyell and All Comers. London; Longman, Brown, Green, Longmans and Roberts

Said erosion increased with slope and decreased in hard rock and in large bodies of water
CHOR 382

Said marine erosion produced plains, but not hills or valleys
CHOR 186

Said valleys lengthen or reduce their upper reaches while building up the lower channels in an attempt to achieve an even gradient throughout
CHOR 382

HALL, JAMES (1811-98)

Contribution to the Geological History of the American Continent, in American Association for the Advancement of Science Proceedings, 31: 29-71 (1883)

Said all mountains formed only after a long accumulation of sediments and 1) upper layers are crumbled by depression, or 2) sea bottom along continent is depressed by influx of sediments
CLA 325-6

Said downward movement of accumulated sediments would cause an upward movement of a continental plateau
CLA 326

Said the Appalachians formed when the weight of sediments, which accumulated along the sea shore, caused the sea floor to sink and crumble, thus raising the mountains
D&B 121; FAIR 55

HITCHCOCK, EDWARD (1793-1864)

Illustrations of Surface Geology, in Smithsonian Contributions to Knowledge. Washington, D.C.; Smithsonian

Said some drift striae, drift and moraines were result of local glaciers and others were caused by a great ice sheet
HLF 163

JUKES, JOSEPH BEETE (1811-69)

Students´ Manual of Geology
Edinburgh; A. and C. Black

Said erosion was mostly performed by the sea prior to elevation of the land
CHOR 394-5

KJERULF, THEODOR (1825-1888)

Besvarelse af den af det Akademiske Collegium D 23 de Mai 1854 . . . ., in Nyt Magazin für Naturvidenskaberne, 9: 31-88, 394

Said igneous rocks consisted of four chemical groups: 1) acid, 2) neutral, 3) basic, 4) ultra-basic
CROSS 355

OWEN, RICHARD (1810-1890)

Key to the Geology of the Globe
New York; A.S. Barnes and Co.

Said internal forces expanded the earth causing the crust to break up into continents
CDBN 351

Said nucleus of the early earth was a dense cube or spherical tetrahedron, and when the heat of the sun caused the crust to swell, cracks occurred along concentric lines, where coal, metals and other minerals then collected
FOHYAG 357-8

SORBY, HENRY CLIFTON (1826-1908)

On the Microscopical Structure of Crystals, Indicating the Origin of Minerals and Rocks, in Geological Society of London Quarterly Journal, 14: 453-500

Said some rocks were formed by simultaneous igneous and aqueous actions
JUDD 198

THOMSON, JAMES

On the Plasticity of Ice as Manifested in Glaciers, in Royal Society of London Proceedings, 8: 455-58 and also see On Current Theories, etc., in 19: 152-69 (1859)

Said glacier movement is the result of ice melting, flowing and refreezing
ROWL 193

VOLGER, GEORG HEINRICH OTTO (1822-97)

Untersuchungen über das Phänomen der Erdbeben in der Schweiz
Gotha; J. Perthes

Said more earthquakes occurred at night
DAV 125

1857-58

DELESSE, ACHILLE ERNEST OSCAR JOSEPH (1817-81)

Recherches sur l´Origine des Roches, in Société Géologique de France Bulletin, Series 2, 15: 728-81

Classified rocks as 1) igneous (effusive), 2) pseudo-igneous (effusive and intrusive), and 3) non-igneous (plutonic)
HSP 63

Said that magmatic minerals might have several modes of formation and that their order of crystallization was not the inverse of their order of fusibility
HSP 63

Said the characteristics of a rock were determined by its mode of origin and its chemical composition
HSP 2

1858

BRONN, HEINRICH GEORG (1800-62)

Untersuchungen über die Entwicklungs-Gesetze der organischen Welt während der Bildungs-Zeitschrift unserer Erd-Oberfläche
Stuttgart; E. Schweizerbart

Said inherent progression in developing organic life occurs concurrently with progression of external conditions which may alter or destroy species
HOOY 305

COTTA, BERNHARD VON (1808-79)

Geologische Fragen
Freiberg; J.G. Engelhardt

Said earth´s crust was two layers, acid above, basic below, as a result of liquation of primitive crust
HSP 63

Said igneous rock formed when rocks in the basic layer of the crust melted (sometimes reacting with upper acid layer), so composition changed little from age to age
HSP 63

Said no rock escaped change
HSP 8

DARWIN, CHARLES ROBERT (1809-82)

On the Origin of Species by Natural Selection. London; J. Murray (1859)

Said controlled breeding (artificial selection) resulted in more variations than would be found in a wild species
ALB 2 168-4

Said natural selection occurred when a spontaneous variation better adapted a species to survive
ALB-2 168-4

Said that variation in wild species was determined by population and size of territory
ALB-2 168-9

Said that, although the fossil record was incomplete, due to the immense length of time involved, an orderly progression had occurred
ALB-2 168-4; GEI 439; RUD 242

DARWIN, CHARLES ROBERT AND ALFRED RUSSEL WALLACE (1809-82 / 1823-1914)

On the Tendency of Species to Form Varieties; and on the Perpetuation of Varieties and Species by Natural Means of Selection, in Linnean Society of London Journal of Proceedings: Zoology, 3: 45-62 (1859)

Said species diversified and/or improved in response to the impact of natural selection on them
RUD 227

EMMONS, EBENEZER (1794-1863)

Agriculture of the Eastern Counties, in Report of the North Carolina Geological Survey. Raleigh; H.D. Turner

Said physical, as well as chemical, properties of soil were important in studying them
MERR 430

MALLET, ROBERT (1810-81)

Earthquake Catalog of the British Association
London; Taylor and Francis

Said earthquakes occurred mostly along the edges of the oceanic basins
DAV 74

NAUMANN, KARL FRIEDRICH (1797-1873)

Lehrbuch der Geognosie (Second Edition) Leipzig; W. Engelmann

Classed rocks as 1) petrogenous, or 2) deuterogenous (formed from 1), and divided petrogenous into 1) ice, 2) haloid, 3) quartz, 4) silicate, 5) ore, or 6) coal
CROSS 357; HTOD 160

OWEN, DAVID DALE (1807-1860)

First Report of a Geological Reconnoissance of the Northern Counties of Arkansas
Little Rock; Johnson and Yerkes

Said internal heat drove hot vapors and gases surfaceward, into water, resulting in hot springs
FOHYAG 366

ROGERS, HENRY DARWIN (1808-66)

Geology of Pennsylvania
Philadelphia; J.B. Lippincott

Said cleavage and foliation resulted when compressive forces moved along parallel planes
GKES 155

SNIDER-PELLEGRINI, ANTONIO

La Création de Ses Mystères Dévoilés. Paris; A. Franck (1859)

Said the crust of the earth cooled on the outside but remained liquid on the inside, thus causing the violent fracturing associated with the Noachian flood that split the supercontinent, separated the New and Old Worlds, and set the continents adrift
BERK 190; CAR-2 284; CDBN 2; D&B 317; HAL 1-2

SORBY, HENRY CLIFTON (1826-1908)

On the Microscopical Structure of Crystals, Indicating the Origin of Minerals and Rocks, in Geological Society of London Quarterly Journal, 14: 453-500

Detailed the composition and internal arrangement of rocks through the use of a microscope
GEI 465-6; JUDD 193+; P&S 117

TYNDALL, JOHN

On the Physical Phenomena of Glaciers, Part 1: Observations on the Mer-de-gace, in Royal Society of London Transactions, 149: 261-307 (1859)

Said glaciers flowed by a process of fracture and resolidification
ROWL 193

1859

COTTA, BERNHARD VON (1808-79)

Die Lehre von den Erzlagerstatten. Freiberg; J.G. Engelhardt

Said metallic veins were result of solutions or magma rising through the earth, or sublimation, or deep-seated deposition revealed as surface ore bodies when surrounding strata eroded away
HTOD 69-71; TODH 325-6

Said ore deposits differed due to heat and pressure differences
HTOD 69-71

HALL, JAMES (1761-1832)

Palaeontology, in Geological Survey of New York
Albany; Van Benthuysen

Said folding, metamorphism and igneous intrusion occurred when the bottom of a geosyncline was bent
FCG 468

Said fractures and intrusions within structures resulted from external forces
FCG 468

Said mountain ranges form when land detritus settles into shallow water along the coast bearing the along trench down until its lower layers are squeezed sideways into the continent causing uplift
FCG 467-8

Said sediment-filled trenches were lines of strength in the earth's crust
FCG 468

Said the height of a mountain range was proportional to the depth of the sediments buried within it
FCG 467-8

Said the higher a mountain was, the younger and thicker its sediments would be
FCG 467-8

Said the weight of debris accumulating near edges of continents bent the crust down, permitting thicker accumulations which then were folded to form mountains
BEL 41

Description and Figures of the Organic Remains of the Lower Helderberg Group and the Oriskany Sandstone, in Natural History of New York, Part 6, Paleontology, Volume 3, Part 1, Albany; C. Van Benthuysen (1847-94)

Said uplift was a continental event which occurred when the heaviest mass of accumulating strata subsided causing crumbling and consequent uplift of upper layers
DRH 241

HUNT, THOMAS STERRY (1826-92)

Notes on Some Points in Chemical Geology, in American Journal of Science, 30: 133-7 (1860)

Said water soaked sediments when displaced and fused by the earth's inner heat, become igneous rocks
GKES 144

JEITTELES, LUDWIG HEINRICH (1830-83)

Das Erdbeben am 15 January 1858 in der Karpathen und Sudeten in seinen Beziehungen zur Atmosphare, in Mitteilungen der geographischen Gesellschaft, 3: 397-414

Said there was a center of seismic focus
DAV 121

NEWBERRY, JOHN STRONG (1822-92)

Rock Oils of Ohio, in Ohio Agricultural Report for 1859 Columbus/Springfield; State Printers

Said carbonaceous shales released oil at low temperatures and the overlying rock pressure then forced it surfaceward through fractures and into porous sandstones
OWEN 60

PRATT, JOHN HENRY (1811-71)

On the Thickness of the Crust of the Earth, in Philosophical Magazine, 17: 327-32

Said cooling earth had produced fracturing and vertical displacement of crust over less dense solid materials below
NCDAFG 236

PRESTWICH, JOSEPH (1812-96)

Sur la Découverte d´Instruments en Silex Associés à des Restes de Mammifères d´Espèces Perdues dans des Couches Non Remaniées d´une Formation Géologique Récente, in Académie des Sciences, Paris. Comptes Rendus, 49: 634-6, 859

Said man and now extinct animals both lived during the Quaternary
CGPS 279

STODDARD, O.N.

Diluvial Striae or Fragments in Situ, in Silliman Journal (American Journal of Science), 28: 227-8

Said pebbles froze in the ground under glaciers and were striated by the grinding passage of ice
FOHYAG 634

WHITTLESEY, CHARLES (1808-86)

On the Drift Cavities, or ´Potash Kettles´ of Wisconsin, in American Association for the Advancement of Science Proceedings, Thirteenth Meeting: 297-301 (1860)

Said it was possible for glaciers to move over level ground and that would explain all the glacial phenomena
AIGT 37; GIA 165

Said kettle holes were result of the action of glacier ice and not of icebergs
EDEKH 120

1860

DAUBREE, GABRIEL AUGUSTE (1814-96)

Etudes et Expériences Synthétiques sur le Metamorphisme et sur la Formation des Roches Cristallenes, in Académie des Sciences, Paris. Mémoires Présentés par Divers Savants, 17: 468-99

Said there was a difference between contact, regional, and structural metamorphism
HDMF 150; HSP 57

HITCHCOCK, EDWARD (1793-1864)

Report on the Geology of Vermont
Claremont, NH; Claremont Manufacturing Company (1861)

Said each rock formation was characterized by specific fossils
FOHYAG 351

Said foliation, cleavage, joints and the transfer of components within a rock mass was caused by galvanism
FOHYAG 351

Said metamorphism caused by pressure had elongated and flattened pebbles while they were in a plastic state
FOHYAG 351; MERR 510-11

HOOKER, JOSEPH DALTON (1817-1911)

On the Origination and Distribution of Species, in American Journal of Science, 29(1): 1-25

Said that species were derivative and mutable
SCHU 93

JAMIESON, THOMAS FRANCIS (1829-1915)

On the Drift and Rolled Gravel of the North of Scotland, in Geological Society of London Quarterly Journal, 16: 347-71

Said erratics came on glaciers or ice-sheets
CHOR 338

MEDLICOTT, HENRY BENEDICT (1829-1905)

On the Geological Structure and Relations of the Southern Portion of the Himalayan Range between the Rivers Ganges and Ravee, in Geological Survey of India Memoir, Volume 3, Article 4

Said rivers cut their own valleys
CHOR 387-8

Said rivers adjust themselves to variations in structure
CHOR 387-8

Said rivers maintained their course, when uplift occurred, by downcutting
CHOR 387-8

OWEN, RICHARD (1804-92)

Palaeontology, or a Systematic Summary of Extinct Animals and Their Relations
Edinburgh; A. and C. Black

Said evolution was the result of a continuous natural creational power
RUD 241

PHILLIPS, JOHN (1800-74)

Anniversary Address of the President, in Geological Society of London Quarterly Journal, 16: xxii+ (varied pages)

Said fossils could identify the epoch (relative time) but not the period (actual time) when a strata formed
CONK 204

RICHTOFEN, FERDINAND PAUL WILHELM VON (1833-1905)

Geognostiche Beschreibung der Umgebung von Predazzo, St. Cassian und der Sesser Alpe
Gotha; J. Perthes

Said the South Tyrol was a reef structure
OWEN 27-8

ROGERS, HENRY DARWIN (1808-1866)

On the Distribution and Probable Origin of the Petroleum or Rock-oil of Western Pennsylvania, New York and Ohio, in Royal Philosophical Society of Glasgow Proceedings, 4: 355-9

Said oil and gas occurred during large-scale metamorphism
OWEN 60

Said oil and gas were usually on anticlinal axes
OWEN 60; RHAP 256

THOMASSY, RAYMOND (1810-63)

Géologie Pratique de la Louisiana. New Orleans; the author

Said salt domes were the result of volcanic action which had heated the waters thus concentrating the salt in the uplifted dome
RTP 208

Said sticky, alluvial mud lumps and mineral springs in the delta were the result of volcanic action
RTP 206

TORELL, OTTO MARTIN (1828-1900)

Bidrag Till Spitzbergens Molluskfauna (Contribution to the Mollusc Fauna of Spitzbergen, together with a General Overview of the Relationships of Nature and Earlier Extension of the Arctic Region). N.P.; N.P. (Dissertation)

Said a continental glacier had covered Western Europe
HAUS 7, 98

WALL, GEORGE PARKES

On the Geology of a Part of Venezuela and of Trinidad, in Geological Society of London Quarterly Journal, 16: 460-70

Said petroleum originated from plant matter and asphalt was the result of chemical action on the same matter
OWEN 7, 33

Report on the Geology of Trinidad; Part 1 of the West Indian Survey. London; Longman, Green, Longman and Roberts

Said oil and gas were symptomatic, not result, of mud volcanoes
HIGG 102

1861

ANDREWS, EBENEZER BALDWIN (1821-80)

Rock Oil, Its Geological Relations and Distribution, in American Journal of Science, Second Series, 3: 92-93

Said oil and gas rose through fractured rocks into anticlines and in direct ratio to the number of fissures present
FOHYAG 400; OPG 567; OWEN 10, 62-3; RHAP 256

BELT, THOMAS (1832-78)

Mineral Veins: An Inquiry into Their Origin, Founded on a Study of the Auriferous Quartz Veins of Australia. London; J. Weale

Said gold is injected into the rocks from ore magmas
HTOD 76-77

BUNSEN, ROBERT WILHELM EBERHARD (1811-99)

Ueber die Bildung des Granites, in Zeitschrift der deutschen geologischen Gesellschaft, 13: 61-63

Said magma was a fused solution
FYPPP 36

Said pressure was determinant of final composition of igneous rocks
HSP 62

HITCHCOCK, EDWARD (1793-1864)

On the Conversion of Certain Conglomerates into Talcose and Micaceous Schists and Gneiss, by the Elongation, Flattening, and Metamorphosis of the Pebbles and the Cement, in American Journal of Science, 31: 372-92

Said metamorphism changed the cement of conglomerates to a crystalline or schistose state while retaining the pebbles in an elongated shape and changed stratified rocks into gneiss, granite, syenite, etc.
MERR 510-11

HUNT, THOMAS STERRY (1826-92)

Bitumens and Mineral Oils, in Montreal Gazette, March 1

Said that oil rises from deep within the earth and accumulates in anticlines along lines of folding
OPG 566; OWEN 61-2; RHAP 256

On Some Points in American Geology, in American Journal of Science, 31: 392-414

Said fissures were filled when metallic solutions seeped down from the surface
HTOD 82-3

Said that water is a totally irresistable, sometimes universal, solvent assisted by heat, pressure and other substances
HTOD 82-3

Said the deposition of ore always involved organic agencies
HTOD 82-3

Said subsidence caused by sediment loading uplifted mountains and lateral compression caused by contractions of the crust caused folding
DRH 244

NICOL, JAMES (1810-79)

On the Structure of the North-western Highlands, in Geological Society of London Quarterly Journal, 17: 85-113

Described the structure of an overthrust
PSG 145

ROTH, JUSTUS LUDWIG ADOLPH (1818-92)

Die Gesteinsanalysen in tabellarischer Uebersicht und mit kritischen Erlauterungen
Berlin; W. Hertz

Classified rocks by chemical composition
HSP 34

SIDELL, WILLIAM HENRY (1810-73)

Report on the Physics and Hydraulics of the Mississippi River, in Appendix A, Corps of Topographic Engineers, Professional Paper 4
Washington, D.C.; Government Printing Office

Said clay particles in river water join into larger particles and thus settle faster than would be expected
CHOR 422

THOMSON, JAMES (1822-92)

On Crystallization and Liquefaction as Influenced by Stresses Tending to Change of Form in Crystals, in Proceedings of the Royal Society, 11: 423-81

Said that the stresses which change the forms of crystals in saturated solutions first cause them to dissolve, then to regenerate
ETHP 371

WALLACE, WILLIAM (1843-97)

Laws Which Regulate the Deposition of Lead Ore in Veins
London; E. Stanford

Said veins are formed by descending solutions affected by electrical action in the fissures
HTOD 85

WHITNEY, JOSIAH DWIGHT (1819-96)

Geological Survey of California
San Francisco; Towne and Bacon

Said diatoms were the major source of oil
HAN 555

WINCHELL, ALEXANDER (1824-91)

First Biennial Report of the Progress of the Geological Survey of Michigan ... Lower Peninsula, in Michigan Geological Survey Annual Report
Lansing; Hosmer and Kerr

Said oil is distilled deep in the earth and rises into porous sandstones overhead
OWEN 61; RHAP 256

1862

DANA, JAMES DWIGHT (1813-95)

Manual of Geology. Philadelphia; T. Bliss and Co.

Said glaciers transported debris inland while icebergs deposited it along the coasts and in deep channels
GREG 119; HLF 164

HUNT, THOMAS STERRY (1826-92)

Notes on the History of Petroleum or Rock Oil, in Smithsonian Institution Annual Report for 1861. Washington, D.C.; Government Printing Office

Said that when organic materials decompos hydrocarbon bodies form in situ and throu fissures to the top of anticlines
OHER 99; OWEN 62, 64

JAMIESON, THOMAS FRANCIS (1829-1915)

On the Ice-worn Rocks of Scotland, in Geological Society of London Quarterly Journal, 18: 164-84

Said erratics had been transported by a glacier stream
CGT 25

JUKES, JOSEPH BEETE (1811-69)

Address to the Geological Section of the British Association at Cambridge, in American Journal of Science, Second Series, 34: 439

Said peneplains were the result of marine erosion
GREG 110-12

Said lateral valleys were cut by rivers draining the heights but longitudinal valleys were cut along strike lines in softer rock
GREG 110-12

On the Formation of Some of the River-valleys in the South of Ireland, in Geological Society of London Quarterly Journal, 18: 378-403

Said the sea attempted to level the land, but after uplift, the main erosion was done by rainwater, streams and rivers
CHOR 395-6

Said erosion affected rocks unequally due to the differences in chemical composition and physical structure
CHOR 395-6

Said some river valleys were created by stream erosion after uplift
MHG 201

LESLEY, JOSEPH PETER (1819-1903)

Observations on the Appalachian Region of Southern Virginia, in American Journal of Science, Second Series, 34, Review: 413-15

Said a river was the junction of anticlines and faulted monoclines, that is, a structural phenomenon
GREG 109

MALLET, ROBERT (1810-81)

Great Neapolitan Earthquake of 1857; Volume 1-2. London; Chapman and Hall

Said Italian earthquakes were caused by volcanic action
WALL 38

MURRAY, DAVID (1830-1905)

Petroleum, Its History and Properties, in Transactions of the Albany Institute, 4: 148-66 (1858-64)

Said petroleum occurs in fractures, porous rocks, and shales
OWEN 66

NEWBERRY, JOHN STRONG (1822-92)

Geological Report, in Report upon the Colorado River of the West, Part III, Edited by J.C. Ives Washington, D.C.; Government Printing Office

Said the canyons had been cut by the waters of the river
GREG 110

RAMSAY, ANDREW CROMBIE (1814-91)

Excavation of Valleys of the Alps, in Philosophical Magazine, Fourth Series, 24 : 377-80

Said running water had cut valleys away
CHOR 401

On the Glacial Origin of Certain Lakes in Switzerland, the Black Forest, Great Britain, North America, and Elsewhere, in Geological Society of London Quarterly Journal, 18: 185-204 and in American Journal of Science, 35: 324-45 (1863)

Said ice sheets had covered northern Europe and dug many present-day lakes
CHOR 336

Said glaciers deepened valleys changing their shape and creating lakes
GREG 125

THOMSON, WILLIAM (LORD KELVIN) (1824-1907)

On the Age of the Sun's Heat, in Macmillan's Magazine, March 1862: 388-93

Said the sun was cooling, and would burn out eventually
CLOUD 43

On the Rigidity of the Earth, in Royal Philosophical Society of Proceedings, 5: 169-70

Said the earth was solid and inflexible, more rigid than glass and nearly as rigid as steel
BRUSH 708; NCDAFG 239

On the Secular Cooling of the Earth, in Royal Society of Edinburgh Transactions, 23: 157-69 (1864)

Said the processes of nature were energy-dissipative and continually draining the earth
HUBB 246

Said the crust had hardened about 100 million years ago
MHG 201

Said the earth solidified 20 to 400 million years ago
BAD 158; CLOUD 42; HUBB 239

Said the sun had burned less than five hundred million years and probably no more than one hundred million years
CLOUD 43

WHITNEY, JOSIAH DWIGHT (1819-96)

Report of the Geological Survey of Wisconsin. Albany; By the State of Wisconsin

Said lead veins formed when joints filled with descending solutions of mineral bearing ocean water in which organic materials decomposed to liberate sulphurated hydrogen gas
FOHYAG 355

1862-63

PRESTWICH, JOSEPH (1812-96)

Theoretical Considerations on the Conditions under Which the Drift Deposits Containing the Remains of Extinct Mammalia and Flint Implements were Accumulated and on Their Geological Age (see also: On the Loess of the Valleys of the Somme and the Seine), in Proceedings of the Royal Society of London, 12: 38-52, 170-3

Said gradual downcutting by a river of its valley could produce terraces
CHOR 337, 448

Said many inland gravels were local, not marine, in origin
CHOR 337, 448

1862-67

GAUDRY, ALBERT (1827-1908)

Animaux Fossiles et Géologie de l'Attique d'après les Recherches Faites en 1855-56 et 1860 sous les Auspices de l'Académie des Sciences. Paris; F. Savy

Said evolution also occurred within species
RUD 245-9

Said evolution was demonstrated by fossil forms and drew phylogenetic trees to show intermediate forms
RUD 245-9

Said that natural selection was too undependable a factor to account for evolution
RUD 245-9

63

NA, JAMES DWIGHT (1813-95)

Manual of Geology. Philadelphia; T. Bliss and Co.

Said forces of erosion ground debris finer as it moved farther and farther
CHOR 367-369

Said height and width of terrace deposits were determined by position of river at flood crest
CHOR 367-369

Said rivers consisted of 1) V-shaped torrent portions, and 2) a U-shaped river portion
CHOR 367-369

Said the Polynesian islands, submerged volcanic remnants, occurred in a line
EOSG 275

JPIUT, JULES-JUVENAL (1804-66)

Etudes Théoriques et Pratiques sur le Mouvement des Eaux dans les Canaux Découverts et à travers les Terrains Perméables (Theoretical and Practical Studies on the Flow of Water in Open Channels and through Permeable Terrains. Paris; Dunod

Said the rate of flow of water underground is directly proportional to energy loss and inversely proportional to distance
CHOW 8

ERGUSSON, JAMES (1808-66)

On Recent Changes in the Delta of the Ganges, in Geological Society of London Quarterly Journal, 19: 321-54

Said meanders occurred when a stream´s natural flow was altered
CHOR 422-3

Said the size of a meander curve was related mathematically to the angle of the slope
CHOR 422-3

OGAN, WILLIAM EDMOND ((1798-1875)

Report on the Geology of Canada Ottawa; Canada Geological Survey

Said a glacier had sculpted the Great Lakes Basin
FOHYAG 636

LYELL, CHARLES (1797-1875)

The Geological Evidences of the Antiquity of Man. Philadelphia; George W. Childs

Said appearance of man was unprecedented, but not catastrophic
CLOUD 50

ROGERS, HENRY DARWIN (1808-1866)

Coal and Petroleum, in Harpers Monthly, 27: 259-64

Said petroleum and hydrogenous gases were derived from fermentation and distillation of marine carbonaceous shales heated in the earth's interior and could be predicted by the percent of carbon present
OWEN 65; RHAP 256

SERRES, MARCEL DE (1783-1862)

Traité des Roches Simples et Composées ou de la Classification Géognostique des Roches de Après Leurs Caractères Minéralogiques et l'Epoque de Leur Apparition. Paris; E. LaCroix

Expressed the average composition of the crust of the earth by mineral content
HSP 40

SORBY, HENRY CLIFTON (1826-1908)

On the Direct Correlation of Mechanical and Chemical Forces, in Proceedings of the Royal Society, 12: 538-50

Said salts whose total bulk increased in water would be less soluble under pressure, but if dissolving reduced their bulk, pressure would make them more soluble (the usual case)
ETHP 369

Ueber Kalkstein-Geschiebe mit Eindrucken, in Neues Jahrbuch fur Mineralogie, Geologie und Palaontologie: 801-7

Said dissolution of limestone out of molassic strata was pressure dependent
ETHP 370

1864

DAWSON, JOHN WILSON (1820-99)

On the Fossils of the Laurentian and the Boulder Drift of Canada, in American Journal of Science, 38: 231-9

Said erratics were deposited, and the Great Lakes cut, when New York, New England and the Canadian Plains subsided and were overrun by the sea
FOHYAG 636

EVANS, EVAN WILHELM (1827-74)

On the Action of Oil Wells, in American Journal of Science, Second Series, 38: 159-66

Said fissures, intergranular and solution porosity, and gas expansion contributed to flow of oil
OWEN 70

PERREY, ALEXIS (1807-82)

Theory of Earthquakes, in American Journal of Science, 37: 1-10

Said earthquakes occurred more often in winter because the earth was in perihelion
DAV 57

PRATT, JOHN HENRY (1811-71)

Speculation on the Constitution of the Earth's Crust, in Proceedings of the Royal Society of London, 13: 253-76

Said the earth's crust was most dense under the oceans, less dense under plains, and least dense under mountains
MATH 393-5, 398

TSCHERMAK, GUSTAV (1836-1927)

Chemisch-Mineralogische Studien - I: Die Feldspatgruppe, in Kongeliche Akademie der Wissenschaften Wien Sitszungberichte, 50: 566-613

Said isomorphous mixtures of two or more minerals were a solid solution
YODER 9

WHITTLESEY, CHARLES (1808-86)

On the Fresh-water Glacial Drift of the Northwestern States, in Smithsonian Contributions to Knowledge, Volume 15, Number 3 Washington, D.C.; Smithsonian Institution (1866)

Said that glaciers had carved some bays and rock-bound lakes
GREG 125

1865

GEIKIE, ARCHIBALD (1835-1924)

Scenery of Scotland Viewed in Connexion with Its Physical Geology. London; Macmillan and Co.

Said lands rise from the waters, topography is created by rain and snow, land is worn away and waters cover it again, in an endless cycle
CHOR 407-9

Said peneplains were the result of marine erosion
GREG 111

JAMIESON, THOMAS FRANCIS (1829-1915)

On the History of the Last Geological Changes in Scotland, in Geological Society of London Quarterly Journal, 21: 161-203

Said glacial ice causes subsidence but the land rebounds when the ice melts
DALY 3; GIA 167; WEG 398

LESLEY, JOSEPH PETER (1819-1903)

On Petroleum in the Eastern Coal Field of Kentucky and Records of Borings in Pennsylvania, in American Philosophical Society Proceedings, 10: 38-68, 187-191

Said distillation of oil in the earth´s crust was an ongoing activity
OWEN 79

Said that the amount of oil found in any given location would depend on the abundance of organic matter and the physical properties and fracturing of the rocks, regardless of structure
OWEN 79

LESQUEREUX, LEO (1806-89)

Report on a Geological Survey of the Lands of the Columbus National Petroleum Company
Columbus; Columbus National Petroleum Company

Said that land plants become woody and decompose to form coal, while water plants remain soft and decompose into petroleum
OWEN 77-8

MENDEL, GREGOR JOHANN (1822-84)

Versuche uber Pflanzenhybriden, in Verhandlungen des Naturforschender Vereins zu Brunn, 4: 3-47

Said variations could be inherited
D&B 70

WINCHELL, ALEXANDER (1824-1891)

On the Oil Formation in Michigan and Elsewhere, in American Journal of Science, 39: 350-3

Said oil and gas is trapped under impervious layers and rises and accumulates in anticlinal arches
OWEN 71

866

AGASSIZ, LOUIS (1807-73)

Physical History of the Valley of the Amazons, Volumes I & II, in Atlantic Monthly, 18: 159-69	Said there had been glaciation in the Amazon region during the Pleistocene AIGT 27; GIA 167

BERTHELOT, MARCELIN PIERRE EUGENE (1827-1907)

Sur l´Origine des Carbures et des Combustibles Minéraux, in Annales de Chimie et de Physique, Fourth Series, 9: 481-3	Said mineral fuels were formed from organic materials by action of heat, water, and alkaline metals OHER 96; OWEN 89

BOMBICCI PORTA, LUIGI (1833-1903)

Sulle Associazioni Poligeniche Applicate alla Classificazione dei Sulfuri Minerali, in Memoire. Reale Accademia delle Scienze dell´ Istituto di Bologna, 6: 475-566 (1867)	Said polyhedra have fundamental coordinating role in isomorphism GORT 515

CLOSE, MAXWELL HENRY (1822-1903)

Notes on the General Glaciation of Ireland, in Royal Geological Society of Ireland Journal, 1: 207-42	Said drumlins were a result of ice movement HLF 173-4

DANA, JAMES DWIGHT (1813-95)

Observations on the Origin of Some of the Earth´s Features, in American Journal of Science, 42: 205-11, 252-3	Said subsidence due to sediment loading did not trigger uplift but that the crust, in places, could support mountains FCG 469
	Said seafloor subsided one foot for every foot of sediment received due to tangential stresses on the crust, not due to weight of the sediment AUB 9-10

EVANS, EVAN WILHELM (1827-74)

On the Oil-producing Uplift of West Virginia, in American Journal of Science, Second Series, 42: 334-8	Said oil, being lighter than water, rose through fractures and was trapped in anticlines OWEN 70

Said the amount of oil trapped would depend on the amount of fracturing and whether rhombic blocks were present thus facilitating lateral and vertical movements
OWEN 70

GUMBEL, KARL WILHELM VON (1823-98)

Ueber das Vorkommen von Eozoon im Ost vayerischen Urgebirge, in Kongeliche Akademie der Wissenschaften Munchen Sitszungberichte, 1(19): 1-46

Said some crystalline rocks were sedimentary in origin
BDCD 156

HUNT, THOMAS STERRY (1826-92)

Petroleum, in Geological Survey of Canada Report of Progress from 1863-1866: 233-62

Said abundance of petroleum required 1) strata with correct attitude, 2) sufficient fissures, and 3) surrounding trap rock
OWEN 82

MAW, GEORGE (1832-1912)

Notes on the Comparative Structure of Surfaces Produced by Subaerial and Marine Denudation, in Geological Magazine, 3: 439-51

Said a peneplain was the result of marine erosion
GREG 112

Said a stream channel was smaller than its total valley
CHOR 405-6

Said that as runoff rate and distance downstream increased, valley walls steepened
CHOR 405-6

NEWBERRY, JOHN STRONG (1822-92)

Prospectus of the Indian Creek and Jacob's Knob Coal, Salt, Oil, Etc., Company with a Geological Report on the Lands, in American Journal of Science, Second Series, 41: 284

Said the oil deposits were in a line of anticlinal axes
OWEN 77

PUMPELLY, RAPHAEL (1837-1923)

Geological Researches in China, Japan and Mongolia, in Smithsonian Contributions, Number 202

Said loess was a lacustrine deposit
GREG 123

JBIDGE, R.N.

On the Denudation of South Africa, in Geological Magazine, 3: 88-91

Said denudation was result of heat, wind, precipitation, and erosion by running water
CHOR 388

HALER, NATHANIEL SOUTHGATE (1841-1906)

On the Formation of Mountain Chains, in Boston Society of Natural History Proceedings, 11: 8-15

Said earth was a solid core covered by igneous zone and topped with a hardened crust which responded to heat loss by contracting, and thus uplifting mountain chains
MERR 528

YLOR, ALFRED (1824-84)

Remarks on the Interval of Time Which Has Passed between the Formation of the Upper and Lower Valley-gravels of Part of England and France, in Geological Society of London Quarterly Journal, 22: 463-8

Said the higher terraces had formed when their rivers had crested during an earlier pluvial period
CHOR 448

ERNER, ABRAHAM GOTTLOB (1749-1817)

Festschrift zur Hundertjahrigen Jubiläum der kongeliche Sachs, Bergakademie Freiberg, am 30 Juli 1866. Dresden; K. Hofbuchdruckerei von C.C. Meinhold und Sohne (1866-67)

Said geological change was largely attributable to water
SMAJ 6

IRKEL, FERDINAND (1838-1912)

Lehrbuch der Petrographie
Bonn; A. Marcus

Classified igneous rocks by 1) dominant minerals and 2) texture and age
CIR 45

1867

OQUAND, HENRI (1831-81)

Sur les Gîtes de Petrole de la Valachie et de la Moldavie et sur l'Age des Terrains Qui les Contiennent, in Société Géologique de France Bulletin, Second Series, 24: 505-69

Said chemical reactions deep in the earth formed petroleum which then escaped upward through springs when strata were consolidated
OWEN 91-4, 506-7

Said springs through which petroleum rose to the surface carried salt also, but the coexistence of the two was only a coincidence
OWEN 509-10

GADOLIN, AXEL WILHELM (1828-92)

Mémoire sur la Deduction d'un Seul Principe de Tous les Systèmes Crystallographiques avec Leur Subdivisions Helsingfors; N.P.

Said there were only 32 types of symmetr possible in crystals
RPM 14

HUNGERFORD, EDWARD

Considerations Relating to the Climate of the Glacial Epoch in North America, in American Association for the Advancement of Science Proceedings, 16: 108-11

Said glaciers were local and unrelated t ice cap growth
HLF 165

LYELL, CHARLES (1797-1875)

Principles of Geology (Tenth Edition) London; J. Murray (1867-68)

Deleted his denial of progressive development in living organisms
CHA 339

1867-68

O'CONNELL, PETER PIERCE LYONS

On the Relation of the Fresh-water Floods of Rivers to the Areas and Physical Features of Their Basins, in Minutes and Proceedings of the Institution of Civil Engineers, 27: 204-17

Said a river's discharge was determined by the size of the drainage basin, its soil and slope, and the amount of rainfall
CHOR 420

1868

GEIKIE, ARCHIBALD (1835-1924)

On Modern Denudation, in Geological Society of Glasgow Transactions, 3: 153-90

Said the earth had formed a crust within the previous one hundred million years
BAD 159

GUMBEL, KARL WILHELM VON (1823-98)

Geognostiche Beschreibung des ostbayerischen Brenzgebirges oder des bayerischen und oberpfalzer Waldbebirges. Perthes; Gotha

Said an amorphous mixture precipitated early in earth's formation from the ocean and heat and pressure subsequently changed it into crystalline rocks and minerals

BDCD 156

RICHTOFEN, FERDINAND PAUL WILHELM VON (1833-1905)

Principles of the Natural System of Volcanic Rocks, in California Academy of Sciences Memoir, Volume 1, Part 2: 39-133	Said there was a sequence in eruptives: basalt - andesite - trachyte - rhyolite - basalt HSP 13

THOMSON, WILLIAM (LORD KELVIN) (1824-1907)

On Geological Time, in Geological Society of Glasgow Transactions, 3: 1-28 (1871)	Said seeds and some small insects could have been transported to earth in meteorites ALB-2 184

1869

GOLOVKINSKIY, NIKOLAI ALEKSEEVICH (1834-97)

Materialy dlya Geologii Rossii (Materials on the Geology of Russia). N.P.; N.P.	Said oscillatory movement helped determine the final form of strata BEL 45

LESLEY, JOSEPH PETER (1819-1903)

Notes on a Map Intended to Illustrate Five Types of Earth-surfaces in the United States, between Cincinnati and the Atlantic Seaboard, in American Philosophical Society Transactions, New Series, 13: 305-11	Said chemical dissolution of limestone occurred underground, while wind, sunshine, freeze-thaw, precipitation, streams and chemical solutions accomplished all surface erosion CHOR 410-11

MOORE, CHARLES (1841-81)

Report on Mineral Veins in Carboniferous Limestone and Their Organic Contents, in Report of the British Association for the Advancement of Science, 39: 360-80	Said elements in sea water enter fissures during marine transgressions and are deposited as mineral veins HTOD 86

PHILLIPS, JOHN (1800-74)

Vesuvius. London; Oxford	Said earth's interior consisted of non-flowable lava or more fluid lava enclosed in caverns NCDAFG 240

RUTIMEYER, LUDWIG (1825-95)

Ueber Thal und See Bildung
Basle; C. Schultze

Said that erosion began when mountains uplifted, varying according to intensity and surroundings, working steadily higher and in uneven stages along a stream´s course
CHOR 455-6

WAAGEN, WILHELM HEINRICH (1841-1900)

Ueber die Ansatzstelle der Hafmuskeln beim Nautilus und den Ammoniten, in Palaeontographica, 17: 185-210 (1867-70)

Said the evolutionary trend of an organism could be traced through its fossil record
D&B 69

1870

CATLIN, GEORGE (1796-1872)

The Lifted and Subsided Rocks of America, with Their Influences on the Oceanic, Atmospheric, and Land Currents. London; Trubner Co.

Said that two currents flow under the Americas: 1) from the north under the Rocky Mountains, and 2) from the south under the Andes, where, heated by volcanoes, it feeds the Gulf Stream while the other excavates the Gulf of Mexico
FOHYAG 461-2

PROCTOR, RICHARD ANTHONY (1837-88)

Meteors and Comets: Their Office in the Universe, in Other Worlds Than Ours (Second Edition) New York; D. Appleton and Co.

Said planets formed from assimilation of meteoric matter
MHG 201

1871

ANSTED, DAVID THOMAS (1814-80)

On Some Phenomena of the Weathering of Rocks, Illustrating the Nature and Extent of Sub-aerial Denudation, in Cambridge Philosophical Society Transactions, Volume 2, Part 2: 387-95

Said alternating hot and cold, wet and dry, of weather caused considerable weathering worldwide
CHOR 402-3

DANA, JAMES DWIGHT (1813-95)

On the Connecticut River Valley Glacier, and Other Examples of Glacier Movement along the Valleys of New England, in American Journal of Science, 2: 233-43

Said the New England glaciers had multiple centers and moved radially
HLF 165

1871

On the Quaternary, or Post-Tertiary of the New Haven Region, in *American Journal of Science*, 1: 1-5

Said the glacial period in New England had involved no iceberg action
GREG 119

GILBERT, GROVE KARL (1843-1918)

On Certain Glacial and Post-glacial Phenomena of the Maumee Valley, in *American Journal of Science*, 1: 339-45

Recognized true nature of terminal moraine and explained its formation by the ebbing glacier
GREG 121

KINGSMILL, THOMAS WILLIAM (1837-1910)

The Probable Origin of ´Loess´ in North China and Eastern Asia, in *Geological Society of London Quarterly Journal*, 27: 376-84

Said loess was an effect of marine deposition
GREG 123

MATTHEW, GEORGE FREDERICK (1871)

Surface Geology of New Brunswick, in *American Journal of Science*, 2: 371-2

Said glacial phenomena in the lower provinces of Canada did not involve effects by icebergs
GREG 120

POSEPNY, FRANZ (1836-95)

Studien aus dem Salinargebiete Siebenburgens, Zweite Abteilung in *K. K. Geologische Reich-anstalt, Vienna. Jahrbuch* 21(1): 123-88

Said salt had accumulated locally in a basin which became a salt sea capturing fossils which could determine the age of the resultant dome
OWEN 103

PUMPELLY, RAPHAEL (1837-1923)

Paragenesis and Derivation of Copper and Its Associates on Lake Superior, in *American Journal of Science*, 2: 188-98, 243-58, 347-55

Said some copper deposits were result of replacement of pre-existing minerals
POGS 129; TODH 330

1871-73

STOPPANI, ANTONIO (1824-91)

Corso di Geologia. Milano; G. Bernardoni and G. Brigola

Classed earthquakes as 1) volcanic, 2) perimetric (near volcano, no eruption), or 3) telluric (no volcanoes nearby)
DAV 93

1872

CONTE, JOSEPH LE (1823-1901)

Theory of the Formation of the Great Features of the Earth´s Surface, in American Journal of Science, 4: 345-55, 460-72

Said the earth cooled inward, in layers, with internal stress relieved periodicall by upheaval and mountain building
AUB 10; MERR 558-9

Said sediments accumulated on the sea floor where subsequent pressures caused them to consolidate and uplift as mountains
GKES 157

DAUSSE, FRANCOIS BENJAMIN

Etudes Relatives aux Inundations et à l´Endiguement des Rivières, in Institute de France, Mémoires Présentés par Divers Savants à l´Académie des Sciences, 20: 287-507

Said all landform slopes, as well as stream slopes, sought equilibrium
CHOR 440

LYELL, CHARLES (1797-1875)

Principles of Geology (Eleventh Edition) London; J. Murray

Said drift had been deposited by rivers before they cut into valleys and became channeled
CHOR 338

Said rocks decompose, and are carried by rivers to the ocean basins where new strata are laid down
ALB-2 140

LYMAN, BENJAMIN SMITH (1835-1920)

Topography of the Punjab Oil Region, in American Philosophical Society Transactions, 15: 1-14

Said oil formed in situ from organic materials which decomposed after deposition in rock without regard to structure
OWEN 86-7

MALLET, ROBERT (1810-81)

Volcanic Energy: An Attempt to Develop Its True Origin and Cosmical Relations, in Proceedings of the Royal Society of London, 20: 438-41

Said that during tectonic movements, the resulting heat generated magma
HSP 68

RECLUS, ELISEE (1830-1905)

La Terre
Paris; Hachette

Said a continent was currently forming in the Sunda Archipelago
BERK 190

Said discrete pockets of magma caused volcanoes and, sometimes, earthquakes
BERK 191

Said earth was no less than three hundred fifty million years of age
BERK 189

Said erratics were the result of a recent ice buildup and decline
BERK 192

Said shift in location of continents was accompanied by a shift in winds and ocean currents
BERK 190

Said slight pre-quake vibrations caused animals to act agitated before an earthquake
BERK 191-2

Said South American and African coasts had once been joined
BERK 190

Said the continents were moving in a circle around the globe
BERK 190

Said the moon might cause earthquakes the same way it affected the tides
BERK 191

Said there had been a supercontinent connecting England and Ireland to Spain and France, Greece to Africa, and Europe to America
BERK 189-90

Said volcanic arcs uplifted the land
BERK 190

TIDDEMAN, RICHARD HILL (1842-1917)

On the Evidence for the Ice Sheet in North Lancashire and Adjacent Parts of Yorkshire and Westmoreland, in Geological Society of London Quarterly Journal, 28: 471-91

Said ice-carried rocks were deposited along lines parallel to the striations on the rocks and coincident with the direction of the ice flow
CGT 11

VOGELSANG, HERMANN PETER JOSEPH (1838-74)

Ueber die Systematik der Gesteinslehre und die Entheilung Gemengten Silikatgesteine, in Zeitschrift Deutschen geologischen Gesellschaft, 24: 507-44

Classified rocks according to abundance and 1) mineral and chemical composition, plus texture, 2) mode of origin and age, and 3) regional distribution
CIR 44-6

Said there were petrographical provinces
HSP 12

1873

DANA, JAMES DWIGHT (1813-95)

On Some Results of the Earth´s Contraction from Cooling, Including a Discussion of the Origin of Mountains, and on the Nature of the Earth´s Interior, in American Journal of Science, Third Series, 5-6: 423-43, 474-75

Said uplift generated lateral pressure which could elevate adjacent anticlines
DRH 243

Said that the earth´s inner heat caused metamorphism and crumbling of buried strata, uplift and adherence of the new land to the older continent
DRH 243, 252

Said the present continents and oceans were formed in the early earth´s crust and each reacts differently to the lateral pressures of the contracting earth, thus ocean borders bend downward under sediment loading by land detritus eventually adhering to the continent as a parallel ocean border begins to bend downward
FCG 470-71

Said that the synclinal trough caused by crustal warping filled with sedments as fast as it sank so coastal lands remained level until tilting and uplift occurred
DRH 243, 252

Said crystalline rocks, such as granite, were metamorphic in origin
GKES 146

Said horizontal compression created large depressions which filled with debris and then were folded into mountains
BEL 41

Said regional buckling was cause, rather than effect, of mountain building
D&B 121; GKES 162

Said volcanoes were fueled from deep in the earth and fusion of crust was a minor effect
GKES 146

Said mountain building was either polygenetic or monogenetic, and the latter are either 1) anticlinorial, or 2) geosynclinorial
FOHYAG 472

Said the earth was a solid globe
NCDAFG 242

On the Position and Height of the Elevated Plateau in Which the Glacier of New England, in the Glacier Era, Had Its Origin, in American Journal of Science, 2: 324-30

Said the New England glacier originated between the St. Lawrence River and Hudson Bay
HLF 165

HUNT, THOMAS STERRY (1826-92)

Geognostical History of the Metals, in American Institute of Mining Engineers Transactions, 1: 331-42

Said pressure causes heated aqueous mineral solutions to move into fissures and cavities, thus forming veins
HTOD 81-2

NEWBERRY, JOHN STRONG (1822-92)

On the Gas Wells of Ohio and Pennsylvania, in American Journal of Science, 5: 225-9

Said oil deposits occurred where there was a 1) source bed, 2) reservoir, and 3) cap rock
OHER 84

ORTON, EDWARD (1829-99)

Hamilton County, in Report on the Third Geological District, 1(1): 425-34	Said there had been interglacial warmups when vegetation again grew DTMG 45

POSEPNY, FRANZ (1836-95)

Die Blei- und Golmei-Erzlagerstatten von Raibl in Karnten, in Jahr Geol., 23: 317-424	Said oxidized ores formed by replacement but sulphide ores were deposited in spaces already in the rock TODH 330

SEEBACH, KARL ALBERT LUDWIG VON (1839-80)

Das mitteldeutsche Erdbeben vom 6 Marz 1872. Leipzig; H. Haessel	Mapped the sound area of an earthquake DAV 126

WINCHELL, NEWTON HORACE (1839-1914)

Surface Geology, in Minnesota Geological and Natural History Survey, First Annual Report: 61-62	Said there had been periods of warmup between glaciers during which vegetation again grew DTMG 45

ZIRKEL, FERDINAND (1838-1912)

Die mikroscopische Beschaffenheit der Mineralien und Gesteine Leipzig; W. Englemann	Classified rocks by their mineral composition GRA 39

1874

ALLPORT, SAMUEL (1816-1897)

On the Microscopic Structure and Composition of British Carboniferous Dolerites, in Geological Society of London Quarterly Journal, 30: 529-67	Said the age criterion was not useful for naming rocks HSP 20

BELT, THOMAS (1832-78)

The Naturalist in Nicaragua London; J. Murray	Said sediments were buried, became heated to liquid state, and then flowed through fissures which have repeatedly opened to permit entry of mineral mixes HTOD 79-81

HIND, HENRY YOULE (1823-1908)

The Figure of the Earth in Relation to Geological Inquiry, in Nature, 10: 165-67

Said the earth is not a perfect sphere because of undulations which cause ocean and lake levels to oscillate
MORT 25

MARSH, GEORGE PERKINS (1801-82)

The Earth as Modified by Human Action. London; Sampson, Low and Searle

Said that the earth's resources were finite and deforestation and soil erosion were related
FAIR 187; MATH 305

MARVINE, ARCHIBALD ROBERTSON (1848-76)

Processes of Erosion in the Colorado Front Range, in United States Geological and Geographical Survey (Hayden Survey), Annual Report for 1873: 44-5

Said subaerial denudation was powerful agent in regional planation
MATH 637

ROSSI, MICHELE STEFANO DE (1834-98)

in Bolletino del Vulcanismo Italiano, 1: i

Devised a ten part scale for comparing the intensities of earthquakes
AD 421; DAV 101-2

1875

CONTE, JOSEPH LE (1823-1901)

On Some of the Ancient Glaciers of the Sierras, in American Journal of Science, 10: 126-39 (see also 5: 325-42)

Said that a glacier erodes the rock over which it moves, carving broad valleys and rock-bound lakes
GREG 125-6

CROLL, JACQUES (1821-90)

Climate and Time in Their Geological Relations: A Theory of Secular Changes of the Earth's Climate. London; Boque

Said glaciers were result of changes in the relationship of the earth and sun affecting the polar caps
HLF 169

GIBBS, J. WILLARD (1839-1903)

On the Equilibrium of Heterogeneous Substances, in American Journal of Science, 16: 441-58 (1878)

Said the number of variables that could change without changing the phase depended on the number of components minus the number of possible phases plus two
FAIR 917; YODER 9

GILBERT, GROVE KARL (1843-1918)

Report on the Geology of Portions of Nevada, Utah, California, and Arizona, in United States Geographical Survey West of the 100th Meridian, 3: 17-187. Washington, D.C.; Government Printing Office

Said that basin-and-range topography was caused by block-faulting
CHOR 548-9

LEVY, AUGUSTE MICHEL (1844-1911)

De Quelques Caractères Microscopiques des Roches Anciennes Acides Considerées dans Leur Relation avec l´Age des Eruptions, in Société Géologique de France Bulletin, Series 3, 3: 199-236

Said differentiation of magma depends on volatiles present as does the formation of alkaline magma
HSP 73

Said the differences in the textures of rocks correlated with their relative ages
HSP 20

PERREY, ALEXIS (1807-92)

Sur la Frequence des Tremblements de Terre Relativement à l´Age de la Lune, in Académie des Sciences, Paris. Comptes Rendus, 81: 690-2

Said earthquakes are more frequent 1) at full and new moons; 2) at perigee than at apogee, and 3) when the moon is near the meridian rather than ninety degrees away
DAV 60

PHILLIPS, JOHN ARTHUR (1829-87)

Rocks of the Mining District of Cornwall and Their Relation to Metalliferous Deposits, in Geological Society of London Quarterly Journal, 31: 319-45

Said fissures occurred during solidification of rock and percolating vapors and water subsequently filled them with minerals
HTOD 89-90

POWELL, JOHN WESLEY (1834-1902)

Exploration of the Colorado River of the West (1869-72)
Washington, D.C.; Smithsonian

Classified rivers as having 1) antecedent (cut prior to folding), 2) consequent (cut after folding), or 3) superimposed (cut in earlier rock structures) and 1) transverse or 2) longitudinal, valleys
CHOR 531-3; FOYHAG 479-80

Said an increase in moisture would increase plant growth and retard erosion
CHOR 555

Said fluvial erosion occurs without regard to structure, and is a major force
CHOR 523

Said no stream could erode below base level and only uplift could cause erosion once equilibrium had been reached
CHOR 523, 518

Said rivers adjust to the level of the land without regard to structure
CHOR 521

Said rivers reduce their valleys until they become flat plains
GREG 111

STERNBERG, H.

Untersuchungen ueber Langen- und Querprofil Geschiebe-fuhrender Flusse, in Zeitschrift fuer Bauwesen, 25: 483-506

Said maximum size of load particles depended on velocity and curve of stream´s profile, and continued attrition in size permitted velocity to decrease downstream
CHOR 443-5

Said river beds in alluvium have concave upward longitudinal profiles, descending from the source, and becoming uneven wherever the stream is less mobile than its load
CHOR 443-5

SUESS, EDUARD (1831-1914)

Die Entstehung der Alpen (Origin of the Alps). Wien; W. Braumuller

Said earth movements are due to gravity and adjustment of larger crust to a shrinking subsurface
PSG 139-40

Said igneous rocks folded just like sedimentary rocks as asymmetrical ranges of mountains arose from the folding
MARG 105

Said the Alps were formed by many folds in the earth lying one over the other, i.e., nappe
AD 397

Said the crust of the earth was heterogeneous
BEL 39

Said the earth is a series of passive blocks (horsts) between which is folded the more pliable crust to form arcuate, unsymmetrical mountain chains
PSG 139-40

TOPLEY, WILLIAM (1841-94)

Geology of the Weald. London; H.M.S.O.

Said the sea leveled highlands, the plains then uplifted and valleys were again cut by forces influenced by the hardness and permeability of the rocks, angle of slopes, rate of erosion, and underlying structure
CHOR 411-4

1876

COOLEY, WILLIAM DESBOROUGH (?-1883)

Physical Geography. London; Dulau and Co.

Said rivers wore away their beds, the speed of the water depending on the slope and depth of the bed with the fastest water being at the surface above the deepest part
CHOR 593-4

Said that rivers in the northern hemisphere were deflected to the right
CHOR 593-4

Said wind and weather alone could not destroy hills and plains
CHOR 593-4

DUTTON, CLARENCE EDWARD (1841-1912)

Critical Observations on Theories of the Earth´s Physical Evolution, in American Journal of Science, 2: 142-5

Said heavier strata sink, causing plastic inner-earth to fold and form synclines and mountains
GKES 164-5

GREEN, ALEXANDER HENRY (1832-96)

Geology for Students and General Readers: Physical Geology
London; Daldy, Isbister

Said subaerial erosion was more important than marine erosion
CHOR 593

KING, CLARENCE (1842-1901)

Systematic Geology, in Geological Exploration of the Fortieth Parallel, Volume 1. Washington, D.C.; Government Printing Office

Suggested there were isolated magmatic reservoirs
HSP 63

1876

POWELL, JOHN WESLEY (1834-1902)

Report on the Geology of the Eastern Portion of the Uinta Mountains and a Region of Country Adjacent Thereto. Washington, D.C.; Smithsonian

Said the angle of slope determined rates of erosion and removal of detritus
CHOR 538

Said the erosive power of a stream depended on its height above base level, the hardness of its bed, and the amount of rainfall received
CHOR 537-9

Said the shape of a stream depended on the interrelationship of velocity, specific gravity of suspended particles, and load mass
CHOR 537

WOODWARD, HORACE BOLINGBROKE (1848-1914)

Geology of England and Wales London; Longmans, Green and Co.

Said that subaerial forces were responsible for topography
CHOR 593

1876-77

OCHSENIUS, KARL CHRISTIAN (1830-1906)

Ueber die Salzbildung der egelnischen Mulde, in Deutsche geologische Gesellschaft Zeitschrift, 28: 654-66

Said deposits of salts resulted when a basin was blocked by a sandbar and sea-water flowed in only as evaporation, causing precipitation of salts, permitted
OWEN 153

1877

CONTE, JOSEPH LE (1823-1901)

On Critical Periods in the History of the Earth and Their Relation to Evolution, and on the Quaternary as Such a Period, in American Journal of Science, 14: 99-114

Said earth's history was a succession of long quiet periods interrupted by brief episodes of violent crustal activity and destruction of life
FOHYAG 467

CORNET, FRANCOIS LEOPOLD AND BRIART, ALPHONSE (1834-87 / 1825-98)

Sur le Relief du Sol en Belgique, in Société Géologique de Belgique Annales, 4: 71-115

Said rain and snow destroyed highlands and carried debris down rivers to the ocean, eventually leveling the land unless uplift compensated for the reduction
CHOR 611

DANA, JAMES DWIGHT (1813-95)

Note on the Glacial Era, in American Journal of Science, 113: 79-80

Said a glacier could move without the land uplifting
HLF 165-6

Said that there had not been a polar ice sheet when New England was glaciated
HLF 165-6

GILBERT, GROVE KARL (1843-1918)

Report on the Geology of the Henry Mountains, in United States Geographical and Geological Survey of the Rocky Mountain Region. Washington, D.C.; Government Printing Office

Said all humid landforms were subject to forces of dynamic (channel) equilibrium acting on transportational slopes
C&B 133

Said all laccoliths were large but the size depended on the depth at which the pressure of the rising magma equaled the weight of the overburden thus spreading the fluid out
GKG 230-1

Said changes in slope could influence erosion differently because of differences in vegetation, water availability, mechanical processes, etc.
CHOR 552-6

Said increasing steepness of slopes would cause a corresponding increase in erosion
CHOR 558-60

Said mountain divides were the same shape as their rivers, regardless of original structure and form
CHOR 558-60

Said mushroom shaped intrusions (laccoliths were formed by upward moving magma and lopoliths by emplacement of passive magma
HSP 18; HUNT 27-9

Said rivers sought grade, that point at which neither erosion nor deposition occurred
CHOR 558-60

Said stream channels moved laterally when it was easier than flowing downslope
GKG 234

Said structure was a result of resistance of the surface and erosional forces present
HUNT 27-9

Said that the size of particles which a river could carry would be affected by maximum size of the particles and the capacity of the river
CHOR 552-6

Said that the slope of a river was the main determinant of speed when friction, load and discharge were equal
CHOR 552-60

Said the shape of an area depended on 1) its rocks´ relative resistance to erosion, 2) the increased erosion caused by an increase in slope, and 3) the increase in slope nearer divides
GKG 233

Said there would be few extremes of topography if erosion were equally effective everywhere
CHOR 558-60

HOFER, HANNS VON HEIMHALT (1843-1924)

Das Erdbeben von Bellund am 29 Juni 1876, in Akademie der Wissenschaften Denkschriften, Volume 74, Part 1: 819-56

Said some earthquakes had multiple foci
DAV 133

MARSH, OTHNIEL CHARLES (1831-99)

Introduction and Succession of Vertebrate Life in America, in American Association for the Advancement of Science Proceedings, 26: 211-58

Said structural changes in organisms were an effect of natural selection
DNS 231

MENDELEEV, DMITRI IVANOVICH (1834-1907)

Neftyanaya Promshlennost U Severo-Amerikanskom Shtate Pensil´vanii I Na Kavkaze (The Petroleum Industry in the North American State of Pennsylvania and in the Caucasus). St. Petersburg; N.P.

Said petroleum formed from hydrocarbons in vapors rising through orogenic faults paralleling mountain ranges after water mixed with hot iron carbides which formed deep within the earth
UPG 286-8; OWEN 149

REYER, EDUARD (1849-1914)

Beitrag zur Physik der Eruptionen und der Eruptivgesteine. Vienna; A. Holder

Said each large-scale type of rock structure was a discrete formation
HSP 24

Die Euganeen Bau und Geschichte eines Vulkanes. Vienna; Alfred Holder

Said magma was heterogeneous, had a Schleren structure, and was either 1) acidic or 2) basic (less common)
HSP 64

Said magma was the result of tectonic movements creating local relief and liquifying rock, plus the influence of volatiles
HSP 64

Said rocks which differed from the two magmas in composition had formed in the transition zone between
HSP 64

RICHTOFEN, FERDINAND PAUL WILHELM VON (1833-1905)

China, Ergebnisse Eigner Reisen und darauf gegrundeter Studien. Berlin; D. Reimer

Said that loess was a result of airborne deposition of materials
GREG 123

ROSENBUSCH, HARRY (KARL HARRY FERDINAND) (1836-1914)

Mikroskopische Physiographie der Mineralien und Gesteine (Microscopical Physiography of of the Rockmaking Minerals) Stuttgart; E. Schweizerbart

Classified rocks according to their mineral composition, followed by age and texture, in that order
FYPPP 34; GRA 39

Classified rocks according to their texture, followed by composition and, in the case of effusives, by age
FYPPP 34

Steiger Schiefer und ihre Contact-zone an den Graititen von Barr-Andlau und Hohwald, in Abhandlungen zur Geologischen Spezielkarte von Elsass-Lothringen, Volume 1, Part 2

Classified igneous rock as 1) intrusive, 2) effusive, and 3) dike (variant of number one)
HSP 17

Said intrusive masses were ringed by zones of lessening metamorphism -- 1) spotted slates, 2) spotted mica-schists, and 3) hornfels
HSP 57

1878

ABBOT, HENRY LARCOM (1831-1927)

On the Velocity of Transmission of Earth Waves, in American Journal of Science, 15: 178-84

Said velocity increases with the intensity of the disturbance and decreases as the wave advances
DAV 146

CHAMBERLIN, THOMAS CHROWDER (1843-1928)

Extent and Significance of the Wisconsin Kettle Moraine, in Wisconsin Academy of Sciences, Arts and Letters Transactions, 4: 201-34

Said moraines could be used to date glacial episodes
DTMG 47

CHAPLIN, WINFIELD SCOTT (1847-1918)

An Examination of the Earthquakes Recorded at the Meteorological Observatory, Tokiyo, in Asiatic Society of Japan Transactions, 6: 353-5

Said Japanese earthquakes were unrelated to full moon maxima or perigee or to crossing the meridian
DAV 179

DANA, JAMES DWIGHT (1813-95)

On Some Points in Lithology, in American Journal of Science, Series 3, 16: 335-43, 431-40

Said texture of rock was determined by origin, not age
HSP 29

DARWIN, GEORGE HOWARD (1845-1912)

On the Bodily Tides of Viscous and Semi-elastic Spheroids, and on the Ocean Tides upon a Yielding Nucleus, in Royal Society of London, Philosophical Transactions, 170: 1-35 (1879)

Said the earth was a solid body
NCDAFG 241

DAWSON, JOHN WILLIAM (1820-99)

Acadian Geology. London; Macmillan and Co.

Said erratics had been deposited by local glaciers, drift ice, and currents from the north
FOHYAG 639

Said glaciers occurred locally, not in sheets, and could not ascend hills or move up out of the ocean
MERR 573-4

HATTORI, ICHIZO (1851-?)

Destructive Earthquakes in Japan, in Asiatic Society of Japan Transactions, 6: 249-75

Said severe earthquakes tended to occur in clusters
DAV 178

HEIM, ALBERT (WILHELM) (1849-1937)

Untersuchungen ueber den Mechanismus der Gebirgsbildungen (Inquiry into the Mechanisms of Mountain Building). Basel; B. Schwabe

Said erosion was of prime importance in wearing mountains away
MARG 105

Said igneous rocks fold just like sedimentary rocks during deformation, creating mountain chains
MARG 105

Said rocks subjected to mechanical deformation split along shear planes and fragments shift to relieve pressure resulting in slaty cleavage or plasticity and may then raise mountains
DGGS 46-8

Said solid rock could fold without breaking
PSG 141

Said the Alps consisted of many folds lying on and over each other
AD 397

HOERNES, RUDOLF (1850-1912)

Erdbeben-Studien, in Geologisches Jahrbuch, 28: 387-448

Said earthquakes were of three types: 1) rock-fall, 2) volcanic, or 3) tectonic (the most common because of mountain building)
DAV 136

KING, CLARENCE (1842-1901)

Systematic Geology, in Geological Exploration of the Fortieth Parallel, 1: 459-529

Said the U-shaped valleys of the Uintas Mountains were formed by glaciers
GREG 126

POWELL, JOHN WESLEY (1834-1902)

Report on the Lands of the Arid Regions of the United States Washington, D.C.; Government Printing Office

Said the so-called renewable resources were renewable only if controlled to prevent waste, and thus to preserve and renew them
FAIR 187

1878

.ITTER, AUGUST (1826-1908)

Untersuchungen ueber die Höhe der Atmosphäre und die Constitution Gasformiger Weltkorper, in Annalen der Physik, 5(3): 405-25, 543-58

Said earth loses heat but contraction of the gaseous nucleus may result in temperature increases
NCDAFG 246

Said interior of the earth is so hot that it is gaseous below a depth of one-tenth its radius
BRUSH 708; NCDAFG 246

.879

:OX, EDWARD TRAVERS (1821-1907)

Annual Reports of the Geological Survey of Indiana, 1876-77-78 Indianapolis; State Printers

Said the glaciers ceased due to atmospheric and meteorological changes combined with the depletion of their icy source
FOHYAG 638

)ARWIN, GEORGE HOWARD (1845-1912)

On the Precession of a Viscous Spheroid, and on the Remote History of the Earth, in Royal Society of London, Philosophical Transactions, 170: 447-53

Said if moon-earth system were once more dynamic, the tides would have been more frequent and powerful eroders
CHOR 597

Said the Pacific Basin was a result of the moon being torn from the earth
HAL 2

:RODDECK, ALBRECHT VON (1837-87)

Die Lehre von den Lagerstatten der Erze. Leipzig; Veit and Co.

Classified ore deposits as 1) formed contemporaneously with the parent rock (sedimentary or eruptive) or 2) formed later by filling spaces or altering the rock
TODH 328

MCGEE, WILLIAM JOHN (1853-1912)

Notes on the Surface Geology of a Part of the Mississippi Valley, in Geological Magazine, 6: 353-61, 412-420, 528

Said interglacial periods involved a considerable span of time
DTMG 48

PUMPELLY, RAPHAEL (1837-1923)

The Relation of Secular Rock-disintegration to Loess, Glacial Drift and Rock Basins, in American Journal of Science, Third Series, 17: 133-44

Said loess consisted of air-borne silt, from dried lakes and rivers, and other decayed matter
GREG 123-4

READE, THOMAS MELLARD (1832-1909)

Limestone as an Index of Geological Time, in Proceedings of the Royal Society, 28: 281-3

Said Laurentian to past Pliocene period lasted 600 million years
FOHYAG 654

1880

CONTE, JOSEPH LE (1823-1901)

The Old River-beds of California, in American Journal of Science, Series 3, 19: 176-90

Said abandoned, debris-filled streams were pre-uplift while downcutting ones are attempting to reestablish balance with the increased height of their canyons
CHOR 612

DANA, JAMES DWIGHT (1813-95)

Manual of Geology (Third Edition) Philadelphia; T. Bliss and Co.

Said earth was fluid filled with no solid core
NCDAFG 244

DUTTON, CLARENCE EDWARD (1841-1912)

Report on the Geology of the High Plateaus of Utah, in United States Geographical and Geological Survey of the Rocky Mountain Region. Washington, D.C.; Government Printing Office

Said removal of strata by erosion triggered uplift of equal mass
CHOR 576; FAIR 55

KARPINSKII, ALEKSANDR PETROVICH (1846-1936)

Observations on Sedimentary Formations in European Russia, in Gornoi Zhurnal, Volume 4 Numbers 11-12

Said a basement of crystalline granite gneiss with a cover of young, slightly deformed sedimentary formations would become a platform area
YANS 127

KERR, WASHINGTON CARUTHERS (1827-85)

Origin of Some New Points in the Topography of North Carolina, in American Journal of Science, Third Series, 21: 216-19 (1881)

Said peneplains were formed by glaciers
GREG 111

AGORIO, ALEKSANDR EUGENIEVICH (1852-?)

Vergleichend-petrographische Studien ueber die massigen Gesteine der Krim. Dorpat; Laakmann

Said magma resulted from action of a specific solvent found in many rocks
HSP 38

MCGEE, WILLIAM JOHN (1853-1912)

On Maximus Synchronous Glaciation, in American Association for the Advancement of Science Proceedings, 29: 447-509

Said that the polar regions were never more extensively glaciated than at present
HLF 166

WHITNEY, JOSIAH DWIGHT (1819-96)

Climatic Changes of Later Geological Times: A Discussion Based on Observations Made in the Cordilleras of North America, in Harvard Museum Zoological Memoir, 7(2): 1-394 (1880-82)

Said glaciers were a mixture of ice and water and the latter facilitated their movement
MERR 580-1

Said the glacial period was the result of an increase in precipitation when an increase in temperature caused an increase in evaporation
MERR 580-1

Said the glaciers were confined to the high cordilleras and other erratics had been carried by icebergs
FOHYAG 641

1880-81

WALLACE, ALFRED RUSSELL (1823-1914)

Geological Climates, in Nature, 23: 124-217, 266-7

Said the Cambrian period had begun about 28 million years ago
ALB-2 193

1881

DARWIN, CHARLES ROBERT (1809-82)

The Formation of Vegetable Mould, Through the Action of Worms London; J. Murray

Said air and rain, and rivers and streams, eroded much more land than did the waves
CHOR 604

FISHER, OSMOND (1817-1914)

Physics of the Earth's Crust
London; Macmillan and Co.

Said there were convection currents, rising under oceans and falling under continents, in the earth's fluid interior
HAL 3

FOREL, FRANCOIS ALPHONSE (1841-1912)

Essai sur les Variations Périodiques des Glaciers, in Archives des Sciences Physiques et Naturelles, 6: 5, 39, 448-60

Devised a ten degree scale to measure earthquake intensity
AD 421; DAV 102

MILNE, JOHN AND ANDREW RAMSAY (1850-1913 / 1814-91)

Report of the Committee Appointed for the Purpose of Investigating the Earthquake Phenomena of Japan, in British Association for the Advancement of Science Report, 200-4

Said earthquakes were due to the earth's crust suddenly splitting away
DAV 201

POYNTING, JOHN HENRY (1852-1914)

Change of State: Solid-Liquid, in Philosophical Magazine, 12: 32-48

Said an increase in pressure on ice or water would lower its melting-point more than if pressure was increased on a mix
ETHP 371

ROCHE, EDOUARD ALBERT (1820-83)

Sur l'Etat Intérieur du Globe Terrestre, in Académie des Sciences, Paris. Comptes Rendus, 93: 364-5

Said earth's shape had been determined during a period in which it rotated at a different rate of speed
NCDAFG 235

1882

CHAMBERLIN, THOMAS CHROWDER (1843-1928)

Bearing of Some Recent Determinations on the Correlation of the Eastern and Western Terminal Moraines, in American Journal of Science, 124: 93-7

Said the interglacial period had been as long as, or longer than, the postglacial period
DTMG 49

DAVIS, WILLIAM MORRIS (1850-1934)

Glacial Erosion, in Boston Society of Natural History, Proceedings, 22: 19-58

Said an effect of glaciers was to remove ridges and loose deposits, thus smoothing the landscape, and re-depositing the materials elsewhere
GREG 127

DUTTON, CLARENCE EDWARD (1841-1912)

Tertiary History of the Grand Canyon District, with Atlas, in United States Geological Survey Monographs, Volume 2 Washington, D.C.; Government Printing Office

Said erosion would reduce the land to base-level
CHOR 582

FISHER, OSMOND (1817-1914)

On the Effect upon the Ocean-tides of a Liquid Substratum beneath the Earth's Crust, in Philosophical Magazine, 14(5): 213-25

Said the layer between the earth's core and crust was fluid
NCDAFG 244

On the Physical Cause of the Ocean Basins, in Nature, 25: 243-4

Said the super-continent split up when the moon was torn from the earth and remaining pieces of granite crust floated off
CDBN 351; HAL 2-3

MCGEE, WILLIAM JOHN (1853-1912)

Relations between Geology and Agriculture, in Iowa State Horticultural Society Transactions, 16: 227-40

Said North American ice sheet had come down out of the Laurentides
DTMG 49

MILNE, JOHN AND THOMAS GRAY (1850-1913 / ?)

On Seismic Experiments, in Royal Society of London, Philosophical Transactions, 173: 863-83

Said the speed of earthquake waves increased with the strength of the shock, decreased with distance, was unaffected by small hills, but was stopped by excavations
DAV 188

ORTON, EDWARD (1829-99)

Source of the Bituminous Matter of the Black Shales of Ohio, in American Association for the Advancement of Science Proceedings, 31: 373-84

Said oil might be found on a terrace which resulted when regional dip approached horizontal
OHER 87

RICHTOFEN, FERDINAND PAUL WILHELM VON (1833-1905)

On the Mode of Origin of the Loess, in Geological Magazine, 9: 293-305

Said unconformities were the result of marine erosion when the land subsided
CHOR 598

Said peneplains were the result of marine erosion
GREG 112

SANDBERGER, KARL LUDWIG FRIDOLIN VON (1826-98)

Ueber die Bildung von Erzgangen-mittelst Auslaugung des Neben-gesteins, in Deutsche Geologische Gesellschaft Zeitschrift, 32: 350-70 (1882-85)

Said veins were formed by lateral secretion of metals already in rocks carried by percolating waters into fissures
AD 319; HTOD 92-3; TODH 326

1883

BARLOW, WILLIAM (1845-1934)

Possible Nature of the Internal Symmetry of Crystals, in Nature, 29: 186-8, 205-7, 404 (1883-4)

Classified crystals by 230 possible internal arrangements
DANA 6-7

CHAMBERLIN, THOMAS CHROWDER (1843-1928)

Artesian Wells, in Geology of Wisconsin, 1: 689-701

Said artesian waters originated on one side of a buried basin, traveled down through a porous stratum contained by two non-porous strata, and rose to the surface on the other side
AGN 82

DAVIS, WILLIAM MORRIS (1850-1934)

Origin of Cross-valleys, in Science, 1: 325-7, 356-7

Said water-gaps were the result of streams maintaining their course in spite of uplift
CHOR 624

GILBERT, GROVE KARL (1843-1918)

A Theory of the Earthquakes of the Great Basin, with a Practical Application, in Salt Lake Tribune, Sep. 30 (also in American Journal of Science, Third Series, 27: 49-53, 1884)

Said earthquakes resulted from a paroxysmal release of the crust of the earth related to the uplift of mountains
EOSG 265; WALL 40

Said that when an earthquake occurs, part of the foot-slope goes up, and another part goes down, creating a cliff face and eventually uplifting a mountain
LBS 50; WALL 38

LAPWORTH, CHARLES (1842-1920)

Secret of the Highlands, in Geological Magazine, 10(2): 120-8, 193-9, 344-77

Said simple structures often hide truer, more complex, structures
PSG 145

LOSSEN, KARL AUGUST

Ueber die Anforderung der Geologie an die petrographische Systematik, in Jahrbuch der preussischen geologischen Landesanstalt und Bergakademie zu Berlin

Said mineral composition varied between rocks which had experienced contact as opposed to regional metamorphism
HDMF 150

NEUMAYR, MELCHIOR (1845-1890)

Ueber klimatische Zonen wahrend der Jura und Kreidezeit, in Akademie der Wissenschaften Denkschriften, 47(1): 277-310

Said fossils indicated variations in ancient climates
SCHU 89

ROSSI, MICHELE STEFANO DE (1834-98)

Programma dell´ Osservatorio ed Archivo Centrale Geodinamico Presso il R. Comitato Giologico d´Italia, in Bollettino del Vulcanismo Italiano, 10: 67-8

Revised his ten-part scale for measuring earthquake intensity, taking into consideration seismographs, people´s receptivity and reactions, and geological effects
DAV 102

1884

DAVIS, WILLIAM MORRIS (1850-1934)

Geographic Classification Illustrated by a Study of Plains, Plateaus, and Their Derivatives, in American Association for the Advancement of Science Proceedings, 33: 428-32 (1885)

Said rivers developed through 1) adolescence (plains which give way to valleys and eroding highlands), 2) maturity (greatest relief), and finally, 3) old age (smooth plain and meandering streams, which uplift could transform into adolescence or the sea might overflow)
CHOR 625-6

Said the class of a geographic formation could be determined by 1) erosional history, or 2) structural origin of the feature
CHOR 625

Said the form of a valley will vary with the rate of uplift
CHOR-2 164

Gorges and Waterfalls, in _American Journal of Science_, Series 3, 28: 123-32	Said the gorges and waterfalls of young rivers become the broad valleys and ripples or old rivers CHOR 624-5

GEER, GERALD JAKOB DE (1858-1943)

Om den Skandinaviska Landisens Anda Utbredning. Stockholm; P.A. Norstedt and Soner	Said clay layers near Stockholm were annual varves and that the timing of the glacial retreat could be calculated from them LIND 220-1; P&S 47

GUENTHER, (ADAM WILHELM) SIEGMUND (1848-1923)

Lehrbuch der Geophysik und physikalischen Geographie Stuttgart; F. Enke (1884-85)	Said temperature and pressure changes caused the earth to undergo transition from solid crust, through plastic, then viscous, then liquid, then gaseous zones down to the core of supercritical gas BRUSH 708

GUTHRIE, FREDERICK (1833-86)

On Eutexia, in _Philosophical Magazine_, Series 5, 17: 462-82	Devised a eutectic classification for substances HSP 37 Said magmas and magmatic rocks were eutectic YODER 9

HOFF, JACOBUS HENDRICUS VAN´T (1852-1911)

Etudes de Dynamique Chimique Amsterdam; Frederik Muller	Divided equilibria into 1) physical and 2) chemical characteristics and described phase incompatibility EUG 481; YODER 9

LEHMANN, JOHANN GOTTLOB (1719-67)

Entstehung der altcrystallinischen Schefergesteine Bonn; Hochgurtel	Said igneous rocks formed at high temperature and pressure which could then change them to schists PSG 144-5

PECKHAM, STEPHEN FARNUM (1839-1918)

Origin of Bitumins, in _American Journal of Science_, 28: 105-17	Said oil in the Appalachians formed from plant remains, but oil in Texas and California formed from animal matter OHER 102

TAIT, PETER GUTHRIE (1831-1901)

Heat. London; Macmillan

Said the earth had a solid core sufficient to keep the globe rigid, a potentially fluid middle layer, and an outer crust kept stable by the solid core
NCDAFG 242

1885

ASHBURNER, CHARLES ALBERT (1854-89)

Geology of Natural Gas, in Science, 6: 42-3

Said the amount of oil found in rock was influenced by 1) porosity and homogeneity of the sandstone, 2) amount of fracturing of the rock, 3) degree of dip and relation of synclines to anticlines, 4) quantity of water, oil and gas present, and 5) pressure
OHER 86

HEIM, ALBERT (WILHELM) (1849-1937)

Handbuch der Gletscherkunde, in Bibliothek geographischer Handbucher Series. Stuttgart; J. Engelhorn

Said rock debris driven forward by glaciers dug out the ground beneath it
MARG 109

JOHNSTON-LAVIS, HENRY JAMES (1856-1914)

Monograph of the Earthquakes of Ischia. London; Dulau and Co.

Said earthquakes of volcanic origin had a small depth of focus
DAV 86

LAPPARENT, ALBERT AUGUSTE COCHOU DE (1839-1908)

Paleontologic Method: Divisions of the Sedimentary Series, in Traité de Geologie. Paris; Savy

Said the oceans had been connected at various times so strata could be correlated by studying their lateral continuity
CONK 205-6

SUESS, EDUARD (1831-1914)

Die Antlitz der Erde (The Face of the Earth). Prag; F. Tempsky (1885-1909)

Said land masses were being forced equatorward by the increasing oblateness of the poles
BLA 68

Said the crust collapses as the earth shrinks and water flows into the fissures, causing sea level to fall
WEG 398

Said the earth was still solidifying and contracting so lighter materials rose to form granite and metamorphic rocks and mountains uplifted as ocean basins sank
HAL 7-8

WHITE, ISRAEL CHARLES (1848-1927)

Anticlinal Theory of Gas Accumulation, in Science, 5: 521-2

Said great gas wells are found in narrow beds on or near the crests of anticlinal axes, though lesser beds may occur down-slope in presence of sufficient dip
OHER 85

1886

BERTRAND, MARCEL ALEXANDRE (1847-1907)

La Chaîne des Alpes et la Formation du Continent Européen, in Société Géologique de France Bulletin, Series 3, 15 : 423-47 (1886-87)

Classified folded zones by their geological age
BEL 40

CONTE, JOSEPH LE (1823-1901)

A Post-Tertiary Elevation of the Sierra Nevada Shown by the River Beds, in American Journal of Science, Series 3, 32: 167-81

Said topography can be correlated with uplift and alternating periods of quiescence
CHOR 613

DARWIN, GEORGE HOWARD (1845-1912)

Geological Time, in British Association for the Advancement of Science Report: 511-8

Said life could be traced back for 100,000,000 years
FOHYAG 656

DAVIS, WILLIAM MORRIS (1850-1934)

Structure of the Triassic Formation of the Connecticut Valley, in American Journal of Science, Series 3, 32: 342-52

Said land could eventually be worn down to an even surface
CHOR 627

EMMONS, SAMUEL FRANKLIN (1841-1911)

Geology and Mining Industry of Leadville, Colorado, in United States Geological Survey Monograph Number 12. Washington, D.C.; Government Printing Office

Said deposits were result of replacement of country rock by solutions which had leached metals from subsurface eruptives
HTOD 88

Notes on Some Colorado Ore-deposits, in Colorado Scientific Society Proceedings, Volume 2, Part 2: 85-105 (1885)

Differentiated between altered (oxidized) minerals and the original (sulphide) ore
TODH 332

FISHER, OSMOND (1817-1914)

An Attempt to Explain Certain Geological Phenomena by the Application to a Liquid Substratum of Henry's Law of the Absorption of Gases, in Cambridge Philosophical Society Proceedings, 6: 19-24

Said earth had a twenty-five mile thick crust which rode on a liquid substratum whose convective currents transported ocean crust to continental crustal areas
NCDAFG 245

On the Variations of Gravity at Certain Stations of the Indian Arc of the Meridian in Relation to Their Bearing upon the Constitution of the Earth's Crust, in Philosophical Magazine, 22(5): 1-29

Said the layer between the earth's core and crust was plastic, thus allowing the crust to shift
NCDAFG 243

JUDD, JOHN WESLEY (1840-1916)

On the Gabbros, Dolerites, and Basalts of Tertiary Age, in Scotland and Ireland, in Geological Society of London Quarterly Journal, 42: 49-97

Said the petrographical province of a rock encompassed its mineralogical composition, the region of occurrence, and the age of the eruption
HSP 22

LEWIS, HENRY CARVILL (1853-88)

Comparative Studies upon the Glaciation of North America, Great Britain and Ireland, in American Journal of Science, 32: 433-8

Said glaciers had originated from both the east and west sides of Hudson Bay
HLF 166

MILNE, JOHN (1850-1913)

On a Seismic Survey Made in Tokio in 1884 and 1885, in Transactions of the Seismological Society of Japan, 10: 1-36 (1887)

Said speed of shockwaves differed within three or four feet and that acceleration increased in soft ground
DAV 188

READE, THOMAS MELLARD (1832-1909)

Origin of Mountain Ranges
London; Taylor and Francis

Said the earth had a solid core which kept the planet rigid and the crust stable, and a potentially fluid layer between the two
NCDAFG 242

RICHTOFEN, FERDINAND PAUL WILHELM VON (1833-1905)

Führer für Forschungsreisende
Berlin; R. Oppenheim

Classified mountains as 1) tectonic (block or fold), 2) trunk or abraded, 3) eruptive, 4) accumulated, 5) plateau, or 6) eroded
CHOR 601

SUESS, EDUARD (1831-1914)

Das Antlitz der Erde (The Face of the Earth). Prag; F. Tempsky (1885-1909)

Classified coastlines, relative to mountain building, as 1) Pacific types, with coastline parallel to the folding, or 2) Atlantic type, with coastline across the folding
HSP 22

1887

DAUBREE, GABRIEL AUGUSTE (1814-96)

Les Eaux Souterraines à l´Epoque Actuelle and Les Eaux Souterraines aux Epoques Anciennes
Paris; V.C. Dunod

Said subterranean water, the major contributor to regional metamorphism and to subsurface ore deposition, was rain, snow, etc., which seeped through the crust plus some liquids freed during igneous intrusion
HTOD 73

LAGORIO, ALEKSANDR EUGENIEVICH (1852-?)

Ueber die Natur der Glasbasis, sowie der Krystallisationsvorgange in eruptiven Magma, in Mineralogische und petrographische Mittheilungen, 8: 421-529

Said magma was a fused solution
FYPPP 36

LAPPARENT, ALBERT AUGUSTE COCHOU DE (1839-1908)

Conference sur le Sens des Movements de l´Ecorce Terrestre in Bulletin de la Société Géologique de France, 15: 215-41

Said horsts were large areas of uplift
PSG 140

ROSENBUSCH, HARRY (KARL HARRY FERDINAND) (1836-1914)

Mikroskopische Physiographie der Mineralien und Gesteine (Second Edition) (Microscopical Physiography of the Rock-making Minerals). Stuttgart; E. Schweizerbart

Based rock classification on texture and mineral composition, in that order, and on mode of occurrence for igneous rocks, with age being a primary consideration for effusives
CIR 46; FYPPP 34

888

HAMBERLIN, THOMAS CHROWDER (1843-1928)

Some additional evidence bearing on the interval between the glacial epochs, in Geological Society of America Bulletin, 1: 469-80 (1890)

Said there had been a very long interglacial period, during which high terraces were cut in river valleys
DTMG 55

AVIS, WILLIAM MORRIS (1850-1934)

Geographic Methods in Geologic Investigation, in National Geographic, 1: 11-26

Said landscape passed through erosional stages, 1) brief youth (high land, increasing relief, torrential currents), 2) maturity (strong and varied relief), a transition period, and 3) indefinitely long old age (feeble current, peneplain)
CHOR-2 170-1, 180-1; PYNE 172

Said mountains could be dated from their form, rocks, and old base levels
CHOR 637

HOLDEN, EDWARD SINGLETON (1846-1914)

Note on Earthquake Intensity in San Francisco, in American Journal of Science, 35: 427-31

Said seismic intensity could be measured on an absolute scale and devised one based on the Rossi-Forel scale of one to nine
DAV 147

MINSHALL, F.W.

History and Development of the Macksburg Oil-field, in Ohio Geological Survey Report, 6: 443-75

Said that oil and gas would be found in domes and dry wells would occur in depressions nearby
OHER 87

NOE, GASTON DE LA (1836-1902)

Formes du Terrain. Paris; Imprimerie Nationale

Said runoff was major determinant of the angle of slopes which decreases over time to nearly level depending on composition and steepness of the surface, the size of the debris and the amount of precipitation
CHOR 627

Said the profile of a river is dependent on time, and independent of load and bed, and would eventually achieve base level
CHOR 630-1

Said angular unconformities were caused by wave action and suberial erosion
CHOR 633

OCHSENIUS, KARL CHRISTIAN (1830-1906)

On the Formation of Rock-salt Beds and Mother-liquor Salts
Berlin; E. Smittler and Sohn

Said petroleum resulted when an inflow of mother-liquors sealed off from the air an organic marine bed containing chloride of aluminum
OWEN 153

OLDHAM, RICHARD DIXON (1858-1936)

On the Law That Governs the Action of Flowing Streams, in Geological Society of London Quarterly Journal, 44: 733-9

Said stream will adopt a velocity sufficient to carry its load but may be affected by grade and/or shape
CHOR 609-10

SUESS, EDWARD (1831-1914)

Das Antlitz der Erde (The Face of the Earth). Prag; F. Tempsky (1885-1909)

Said the Noachian flood was local to the lower Euphrates valley and was caused by an earthquake in the Persian Gulf, probably aided by a cyclone out of the g
WOOD 87-8

TEALL, JETHRO JUSTINIAN HARRIS (1849-1924)

British Petrography: With Special References to the Igneous Rocks. London; Dulau and Co.

Said some substances concentrate in the cooler parts of a solution before crystallization occurs
HSP 73

Classified igneous rocks according to composition and texture
HSP 29

Said graphic granite was eutectic
HSP 37

TRYBOM, FILIP

Bottonprof Från Svenska Insjöar, in Geologiska Föreningen I Stockholm Förhandlingar, 10: 489-511

Used pollen grains as index fossils
HCMP 280

1889

BALLORE, FERNAND JEAN BAPTISTE MARIE BERNARD MONTESSUS DE (1851-1923)

Sur la Répartition Horaire des Seismes et Leur Relation Supposée avec les Culminations de la Lune, in Académie des Sciences, Paris. Comptes Rendus, 109: 327-30, 392

Said earthquake frequency did not vary throughout the day
DAV 172

BUCKMAN, SYDNEY SAVORY (1860-1929)

On the Cotteswold, Midford, and Yeovil Sands and the Division between Lias and Oolite, in Geological Society of London Quarterly Journal, 45: 440-73

Said that portions of continuous paleontological zones of uniform lithology could be of differing ages
WSSB 226

DAVIS, WILLIAM MORRIS (1850-1934)

Rivers and Valleys of Pennsylvania, in National Geographic, 1: 183-253

Said maturity lasted a long time in the normal cycle of a river
CHOR 638-40

Said rivers have a cycle of youth (imperfect drainage, lakes and ponds, lots of evaporation, few channels, not much wastage), maturity (extensive drainage system, few lakes and ponds, little evaporation, lots of wastage), and old age
CHOR-2 180-1

Topographic Development of the Triassic Formation of the Connecticut Valley, in American Journal of Science, Series 3, 37: 423-34

Said topography could form in cycles (polygenetically) with two planes existing where a higher peneplain rose above a later one
CHOR 636

DUTTON, CLARENCE EDWARD (1841-1912)

On Some of the Problems of Physical Geology, in Philosophical Society of Washington Bulletin, 11: 51-64

Said that the surface elevation of the earth balanced the density and thickness of the crust
WOOD 252

Said isostasy was a slow, rather than simultaneous, response to loading changes and that smaller structures might not trigger adjustment at all
DALY 4

LEVY, AUGUSTE MICHEL (1844-1911)

Structures et Classification des Roches Eruptives. Paris; Librairie Polytechnique, Baudry et Cie	Said igneous rocks should be classed solely by texture and mineralogical composition HSP 17-8

LOEWINSON-LESSING, FRANTS IULEVICH (1861-1939)

Note sur la Structure des Roches Eruptives, in Société Belge de Géologie Bulletin, 3: 393-8	Said a formation was a genetic unit determined by the dominant rock type HSP 25

REBEUR-PASCHWITZ, ERNST VON (1861-95)

Earthquake of Tokio, April 18, 1889, in Nature, 40: 294-5	Said seismic waves detected in Germany had traveled through earth from Tokyo BRUSH 709

WRIGHT, GEORGE FREDERICK (1838-1921)

The Ice Age in North America New York; D. Appleton and Co.	Said glaciation in North America, which occurred because the continent was so high, had lasted only a short time but involved several pulses of the glacier as it retreated DTMG 53-4

1890

BROGGER, WALDEMAR CHRISTOFER (1851-1940)

Mineralien der Syenit-Pegmatit-Gange der Sudnorwegischen Augit und Nephelinsyenite, in Zeitschrift für Kristallographie und Mineralogie, 16: 1-663	Said differentiation occurred in the same sequence as crystallization HSP 13 Said rocks with common parentage could have regional variances RPS 231

DUTTON, CLARENCE EDWARD (1841-1912)

Charleston Earthquake of 31 August 1886, in United States Geological Survey Report for 1887-88: 203-528 Washington, D.C.; Government Printing Office	Said earthquakes could have more than one epicenter DAV 149-151

FOUQUE, FERDINAND ANDRE AND LEVY, AUGUSTE MICHEL (1828-1904 / 1844-1911)

Note sur la Structure des Roches Eruptives, in Société Belge de Géologie de Paléontologie et d'Hydrologie Bulletin, 4: 144-50	Classified igneous rocks by texture and mineralogical makeup HSP 17-8

1890

GILBERT, GROVE KARL (1843-1918)

Lake Bonneville, in United States Geological Survey Monograph Number 1. Washington, D.C.; Government Printing Office — Said spits, bars and barrier beaches were like features differing only in the processes creating them
C&B 135

LOEWINSON-LESSING, FRANTS IULEVICH (1861-1939)

Etude sur la Composition Chimique des Roches Eruptives, in EEND (Mem.): 221-35 — Classified rocks according to their chemical composition
GRA 39

WALCOTT, CHARLES DOOLITTLE (1850-1927)

Geologic Time as Indicated by the Sedimentary Rocks of North America, in American Association for the Advancement of Science Proceedings: 129-69 — Said geologic time totaled 55,140,000 years through five periods: Cenozoic, 2,900,000 years; Mesozoic, 7,240,000 years; Paleozoic, 17,500,000 years; Algonkian, 17,500,000 years; and Archean, 10,000,000 years
FOHYAG 657-8

1891

SCHOENFLIES, ARTHUR MORITZ (1853-1928)

Krystallsysteme und Krystallstructur. Leipzig; B.G. Teubner — Said the internal symmetry of crystals involved 230 different arrangements
DANA 6-7; PHIL 234

1892

CHATELIER, HENRI-LOUIS LE (1950-1936)

Sur la Théorie du Regel, in Académie des Sciences, Paris. Comptes Rendus, 114: 62-4 — Said an increase in pressure on a solid in a liquid would increase solubility and fusibility
ETHP 371

DAKYNS, JOHN ROCHE AND TEALL, JETHRO JUSTINIAN HARRIS (1832-96 / 1849-1924)

On the Plutonic Rocks of Garabel Hill and Meall Breac, in Geological Society of London Quarterly Journal, 48: 104-20 — Said igneous rocks rarely phased into other kinds of igneous rocks
FYPPP 44

GILBERT, GROVE KARL (1843-1918)

The Moon's Face: A Study of the Origin of Its Features, in Philosophical Society of Washington Bulletin, 12: 241-92 — Said lunar craters were formed by impacts not by volcanoes
BAZ 69

IDDINGS, JOSEPH PAXSON (1857-1920)

On the Crystallization of Igneous Rocks, in Philosophical Society of Washington Bulletin, 11: 65-113 and 12: 89-213

Classified rocks according to their chemical composition and said regional rocks might vary in spite of a common parent magma
HSP 22

Said igneous rocks were alkali or subalkali, and formed from a common magma solution which differentiated according to proportions of oxides present during solidification
HSP 22, 65

NEWCOMB, SIMON (1835-1909)

On the Law and the Period of the Variation of Terrestrial Latitudes, in Astronomische Nachrichten, 130: 1-6

Said the solid earth is slightly more rigid than steel
BRUSH 709

SALISBURY, ROLIN D. (1858-1922)

Distinct Glacial Epochs and the Criteria for Their Recognition, in Journal of Geology, 1: 61-84 (1893)

Defined a glacial epoch as glacial ice readvancing after it had withdrawn to or above the Canadian border as defined by twelve criteria (forest beds, animal remains, etc.)
DTMG 57-8

1893

BALLORE, FERNAND JEAN BAPTISTE MARIE BERNARD MONTESSUS DE (1851-1923)

Quoted in the Royal Society of London Philosophical Transactions, A:1115-6 by Mr. C. Davison

Said that the frequency of earthquakes did not vary from season to season
DAV 170-1

BUCKMAN, SYDNEY SAVORY (1860-1929)

The Bajocian of the Sherbourne District; Its Relation to Subadjacent and Suprajacent Strata, in Geological Society of London Quarterly Journal, 49: 479-521

Said paleontological zones were divided into hemeras, the times when a given species was deposited
CONK 206; WSSB 226

KING, CLARENCE (1842-1901)

Age of the Earth, in American Journal of Science, Third Series, 145: 1-20, 45-51

Said earth was solid to one-fifth of its radius
NCDAFG 245

Said thermal properties of rocks indicated 24 million years as the maximum age of the earth
HUBB 242

Said the earth was twenty to thirty million years old
CGPS 278

KOTO, BUNDJIRO (1856-1935)

On the Cause of the Great Earthquake in Central Japan, 1891, in Tokyo University College of Science Journal, 5: 297-353

Said earthquakes were result of fault displacement which caused land to move both vertically and horizontally
EOSG 265; WALL 38

MCGEE, WILLIAM JOHN (1853-1912)

Note on the "Age of the Earth", in Science, 21: 309-10

Said the earth´s age was ten million to five trillion years
ALB-2 192-4

POSEPNY, FRANZ (1836-95)

Genesis of Ore Deposits, in American Institute of Mining Engineers Transactions, 23: 197+ (1894) and 24: 924+ (1895)

Said most ore deposits postdated the parent rock and ore deposition occurred in earth´s outer aqueous layer by lateral secretion or descending solutions, and occurred in deeper layers filled by core minerals in ascending hot waters
POGS 125; HTOD 94-6; TODH 331

READE, THOMAS MELLARD (1832-1909)

Measurement of Geological Time, in Geological Magazine, 10: 97-100

Said Cambrian was 95,040,000 years ago
FOHYAG 658

VOGT, JOHAN HERMAN LIE (1858-1932)

Bildung von Erzlagerstatten durch Differentiationsprocesse in basischem Eruptivmagmata, in Zeitschrift fur praktische Geologie, 1: 4-11, 125-43, 257-84

Said ore deposits were a product of differention in upswelling basic eruptive magma
HTOD 98-9; TODH 332

WIIK, FREDRICK JOHAN (1839-1909)

Utkast Till Ett Kristallokemiskt Mineral System (Outline for a Crystal-Chemical Mineral System). Helsingforsiae; Societatis Litterariae Fennicae

Designed a classification system based on three fundamental elements, one positive, one negative and one neutral, which determined the molecular makeup of minerals which, in turn, determined external characteristics
HAUS 25

1893-94

WALTHER, JOHANNES (1860-1937)

Einletung in die Geologie als historische Wissenschaft
Jena; Fischer

Said the Geochemical Cycle was 1) weathering, 2) erosion, 3) transportation, 4) corrasion, 5) deposition, 6) diagenesis, and 7) metamorphosis
BDCD 157

1894

KARPINSKII, ALEKSANDR PETROVICH (1846-1936)

Obshchiy Kharakter Kolebaniy Yemnov Kory v Predelakh Yevropeyskoy Rosii (Sur le Caractère Général des Mouvements de l´Ecorce Terrestre dans la Russie d´Europe) (On the General Character of Oscillations of the Crust of the Earth in European Russia), in Izvestiya Akademii Nauk SSSR Bulletin, 1: 1-19 and Annales de Géographie, 5: 179-192 (1896)

Said the tectonics of a region could be determined by reconstructing its paleogeography
YANS 127

Said oscillations in the crust caused uplift and subsidence
BEL 46

OMORI, FUSAKICHI (1868-1923)

On the Aftershocks of Earthquakes, in Tokyo University College of Science Journal, 7: 111-200 (1895)

Said variations in atmospheric pressure caused diurnal fluctuations in the frequency of earthquakes, with most occurring around one hour after midnight
DAV 223

Described method for calculating the frequency of aftershocks up to eight years after an earthquake
DAV 212

RIECKE, (CARL VICTOR) EDUARD (1845-1915)

Ueber das Gleichgewicht zwischen einem festen homogen deformirten Körper und einer flüssigen Phase insbesondere über die Depression des Schmelzpunctes durch einseitige Spannung, in Nachrichten von der Georg-Augusts Universität und der Kongeliche Gesellschaft der Wissenschaften zu Göttingen: 278-84 (1894)

Said elastic deformation lowers the melting point of a solid under pressure
ETHP 371

UPHAM, WARREN (1850-1934)

Diversity of the Glacial Drift along Its Boundary, in American Journal of Science, 147: 358-65

Said entire glacial period had lasted for a brief time
DTMG 63

VOGT, JOHAN HERMAN LIE (1858-1932)

Ueber die durch pneumatolytische Processe an Granit gebundenen Mineral-Neubildungen, in Zeitschrift fur praktische Geologie, 2: 458-65

Said ore deposits formed from vapors and fluids escaping from cooling magma
POGS 126; TODH 332

WALTHER, JOHANNES (1860-1937)

Lithogenesis der Gegenwart, in his Einleitung in die Geologie als historiche Wissenschaft Jena; G. Fischer (1893-94)

Said that the sediments contain a record of earlier climates
SCHU 89

1895

BALLORE, FERNAND JEAN BAPTISTE MARIE BERNARD MONTESSUS DE (1851-1923)

Relation entre le Relief et la Sismicité, in Archives des Sciences Physiques et Naturelles, 34: 113-33

Said features of greatest relief, above or below sea level, were the most unstable seismic areas
DAV 165

CONTE, JOSEPH LE (1823-1901)

Genesis of Ore Deposits (by Posepny); Discussion, in American Institute of Mining Engineers Transactions, 24: 942-1006

Said ore deposits form from waters which usually are 1) alkaline, 2) hot, 3) under pressure, and 4) ascending through fissures in mountain regions of metamorphic and igneous rock
HTOD 96

DANA, JAMES DWIGHT (1813-95)

Manual of Geology. New York, Cincinnati; American Book Company

Said oceanic volcanoes produced basalt while lighter andesites and trachytes were produced on land so the weight differences kept them in place
DRH 249

Said that compression by the contracting earth in conjunction with sediment loading raised mountains
DRH 244-5

DAVIS, WILLIAM MORRIS (1850-1934)

Development of Certain English Rivers, in Geographical Journal, 5: 127-46

Said streams cut their valleys down toward base level until they achieve grade
CHOR-2 175-9

GILBERT, GROVE KARL (1843-1918)

Niagara Falls and Their History, in Journal of Geology, 3: 121-7

Said the equinoxes were responsible for the rhythmical alternations in sandstone, shale, and limestone
FOHYAG 659

KENNEDY, ROBERT GREIG (1851-1920)

Prevention of Silting in Irrigation Canals, in Minutes and Proceedings of the Institution of Civil Engineers, 119: 281-90

Said flow velocity in an adjusted self-regulating stream was a function of depth
CHOR 608

PEANO, GIUSEPPE (1858-1932)

Sopra lo Spostamento del Polo Sulla Terra, in Reale Accademia Delle Scienze di Torino. Atti, 30: 515-23, 845-52

Said the north and south poles moved around
MEYER 6564

SPURR, JOSIAH EDWARD (1870-1950)

Economic Geology of the Mercur Mining District, Utah, in USGS Annual Report 16, Part 2: 343-455

Said primordial magmatic waters caused Mineralization
POGS 126

1896

COPE, EDWARD DRINKER (1840-97)

The Primary Factors of Organic Evolution. London; Open Court Publishing Co.

Said variation and new species was an effect of changes in environment
DNS 232

HISE, CHARLES RICHARD VAN (1857-1918)

Principles of North American Precambrian Geology, in United States Geological Survey, Sixteenth Annual Report (1): 571-874

Said that Precambrian graphite was organic in origin
CDW 268

LEONHARD, RICHARD

Das mittelschlesische Erdbeben vom 11 Juni 1895 und die schlesischen Erdbeben, in Zeitschrift fur Erdkunde zu Berlin, 31: 1-21

Said an earthquake was caused by movement of orographic blocks
EOSG 265-6

LUGEON, MAURICE (1870-1953)

La Region de la Breche du Chablais, in Service de la Carte Géologique de France de Topographies Souterraines Bulletin, 7 (49): 337-646

Said mountains were rootless and floating
MARG 107

POWELL, JOHN WESLEY (1834-1902)

Physiographic Features, in Physiography of the United States. New York; Geographic Society

Classified mountains as 1) volcanic or 2) diastrophic
CHOR 541

SHIMEK, BOHUMIL (1861-1937)

A Theory of the Loesses, in Iowa Academy of Science Proceedings, 3: 82-6

Said loess was a non-marine deposit of dust trapped by vegetation and topography during interglacial periods
DTMG 67

WIECHERT, JOHANN EMIL (1861-1928)

Ueber die Massenvertheilung im Inneren der Erde, in Deutsche Naturforscher Verhandlungen, 2 (1): 42-3

Said the earth was a thin crust around a shell with an iron core and had a density of about 3.2 for crust, 5.6 for shell, and 8.2 in the core, due to chemical composition differences between rock and iron
BRUSH 705

1897

BECKER, GEORGE FERDINAND (1847-1919)

Some Queries on Rock Differentiation, in American Journal of Science, Series 4, 3: 21-40

Said that magmas were stiff imulsions similar to modeling clay, always present in the crust, but only potentially liquid
HSP 68

CHAMBERLIN, THOMAS CHROWDER (1843-1928)

Group of Hypotheses Bearing on Climatic Changes, in Journal of Geology, 5: 653-83

Said glacial periods were caused by the gradual reduction of temperature resulting from the absorption of carbon dioxide by oceans and the weathering of newly uplifted rocks
CONK 262; FENT 290-1

Supplementary Hypothesis Respecting the Origin of the Loess of the Mississippi Valley, in American Journal of Science, 5: 795-802

Said loess was deposited by water
GREG 124

DAWSON, JOHN WILLIAM (1820-99)

Relics of Primeval Life. New York; H. Rewell Co.

Said graphite formed from a bed of vegetable matter
CDW 268-9

GUENTHER, SIEGMUND (1848-1923)

Handbuch der Geophysik und physicalischen Geographie Stuttgart; Ferdinand Enke

Said variations in temperature and pressure at various depths beneath the earth's crust would result in seven layers: 1) crust, 2) plastic zone, 3) magma, 4) fluid, 5) ordinary gas, 6) supercritical gas, and 7) a core of monoatomic gases
NCDAFG 247

LOEWINSON-LESSING, FRANTS IULEVICH (1861-1939)

Studien über die Eruptivgesteine, in Comptes Rendus Seventh International Geological Congress, St. Petersburg: 191-465 (1899)

Said end product of differentiation was a pure magma and distinguished between monotectic, heterotectic, and isotectic magmas
HSP 39

Said magma differentiated in response to assimilation of foreign materials
HSP 73

Classified rocks according to the ratio of oxides they contained
HSP 39

MERCALLI, GIUSEPPE (1850-1914)

I Terremoti della Liguria e del Piemonte. Naples; Lanciano e Pinto

Devised a ten-scale earthquake classification system
DAV 106-7

1897

MILNE, JOHN (1850-1913)

Seismological Investigation, in British Association for the Advancement of Science Report 171-6

Said acceleration in speed of initial tremors could be shown by drawing a velocity curve
DAV 191, 196

Sub-Oceanic Changes, in Geographical Journal, 10: 129-46

Said earthquakes were stronger and more frequent on the margins of the sub-oceanic banks and continental slopes
DAV 200

WALTHER, JOHANNES (1860-1937)

Versuch einer Classification der Gesteine auf Grund der vergleichenden Lithogenie, in Comptes Reudus Seventh International Geological Congress, St. Petersburg: 9-25 (1899)

Classified all rocks based on their origin and history
FYPPP 35

1898

BROGGER, WALDEMAR CHRISTOFER (1851-1940)

Gangefolge des Laurdalits, in Eruptivgesteine des Kristianiagebietes. Norske Videnskaps-Akademi, Oslo. Matematiske-Naturvidenskapelig Klasse. Skrifter. 6: 1-377

Said substances in magma were concentrated by thermal convection currents
HSP 73

CHAMBERLIN, THOMAS CHROWDER (1843-1928)

The Ulterior Basis of Time Divisions and the Classification of Geologic History, in Journal of Geology, 6(5): 445-62

Said the rise and fall of sea level corresponded with the deepening of ocean basins and the leveling of continents
CONK 262

HISE, CHARLES RICHARD VAN (1857-1918)

Metamorphism of Rocks and Rock Flowage, in Geological Society of America Bulletin, 9: 269-328

Said dissolution caused reduction in the direction of greatest stress and increase in that of least stress
ETHP 371

ROSIWAL, AUGUST KARL (1860-1923)

Ueber geometrische Gesteinsanalysen, in Geologische Bundesanstalt, Vienna. Verhandlungen: 143-75

Said geometrical analysis of the composition of minerals should be done linearly
HSP 80

1899

CHAMBERLIN, THOMAS CHROWDER (1843-1928)

An Attempt to Frame a Working Hypothesis of the Cause of Glacial Periods on an Atmospheric Basis, in Journal of Geology, 7: 667-685, 751-787

Said earth had taken form through accretion
D&B 81, 86

DAVIS, WILLIAM MORRIS (1850-1934)

The Geographical Cycle, in Geographical Journal, 14: 481-504

Said erosional activity was a continuation of the chemical action of air and water and the mechanical action of wind, temperature, and precipitation in its various forms
CHOR-2 167

Said the geographic cycle consisted of periods of varying lengths of time, relief, and rates of change
CHOR-2 180-1

Said the winds are arid deserts, as well as the ice of the frigid deserts, effected erosion
CHOR-2 171

Said young rivers cut into their outer banks at every turn, becoming serpentine, and since outward cutting continues after grade has been established, eventually they develop a flood plain
CHOR-2 185-7

JOLY, JOHN (1857-1933)

Estimate of the Geological Age of the Earth, in Royal Dublin Society Scientific Transactions, 7: 23-66 (1902)

Said land had consolidated and water had condensed on the earth 80 to 90 million years earlier judging by measurements of sodium in the sea
BAD 160; FOHYAG 660

LAUNAY, LOUIS DE (1860-1938)

Recherche, Captage et Arménagement des Sources Thermominérales Paris; Baudry

Said metallic deposits formed from warm meteoric waters
POGS 125

OLDHAM, RICHARD DIXON (1858-1936)

On the Propagation of Earthquake Motion to Great Distances, in Royal Society of London. Philosophical Transactions, (A) 194: 135-174

Said earth had core, probably of iron, which is surrounded by magma (stony or glassy)
BRUSH 170

Report on the Great Earthquake of 12th June 1897, in Geological Survey of India Memoir, 29: 1-379

Identified P and S waves occurring during the earthquake and attributed it to movement along a thrust plane deep underground
BRUSH 710; EOSG 266

OMORI, FUSAKICHI (1868-1923)

Notes on the Earthquake Investigation Committee Catalogue of Japanese Earthquakes, in Tokyo University College of Science Journal, 11: 389-437 (1898-99)

Said that the occurrence of small earthquakes varied annually the same as destructive and semi-destructive ones varied
DAV 222

SPRING, WALTHERE (1848-1912)

La Plasticité des Corps Solides et Ses Rapports avec la Formation des Roches, in Académie Royale des Sciences de Belgique. Classes des Sciences Bulletin: 790-815

Said agglutination occurred when compression encountered plasticity
HSP 52-3

THOMSON, WILLIAM (LORD KELVIN) (1824-1907)

The Age of the Earth as an Abode Fitted for Life, in Philosophical Magazine, Series 5, 47: 66-90 and in Science, 9: 665-74, 704-11

Said he agreed that the earth could not be more than twenty-four million years old
CLOUD 43

Said the earth was twenty to, at most, forty million years old
BAD 158

Said there would have been free oxygen in the earth's atmosphere only after solidification (20-25 million years ago), after which plants arose, produced more oxygen (a few hundreds or thousands of years) and then animal life was sustainable
ALB-2 235; HUBB 242

Said the earth was at least twenty to thirty million years old
CGPS 278

WALCOTT, CHARLES DOOLITTLE (1850-1927)

Pre-Cambrian Fossiliferous Formations, in Geological Society of America Bulletin, 10: 199-244

Said graphite appeared to be the result of alteration of coal beds and therefore would be organic in origin
CDW 168

Said the thickness of sedimentary rock could affect the structural features below
CDW 265

LIST OF SOURCES AND THEIR ABBREVIATIONS

AD Adams, Frank D. The Birth and Development of the Geological Sciences. New York: Dover. 1954, c1938.

AGN Agnew, Allen F. G.K. Gilbert and Groundwater, or 'I have drawn this map with great reluctance.' In The Scientific Ideas of G. K. Gilbert, edited by Ellis Y. Yochelson. G. S. A. Special Paper 183: 81-91. 1980.

AIGT Carozzi, Albert V. Agassiz's Influence on Geological Thinking in the Americas. Archives des Sciences, 27(1): 5-38. 1974.

ALB Albritton, Claude C. Philosophy of Geohistory: 1785-1970. Stroudsberg, PA; Dowden, Hutchinson & Ross, Inc. 1975.

ALB-2 -----. The Abyss of Time. San Francisco, CA; Freeman, Cooper. 1980.

ALB-3 -----. The Fabric of Geology, Reading, MA; Addison-Wesley, Inc. 1963.

ALCOCK Alcock, Frederick J. A Century in the History of the Geological Survey of Canada. In National Museum of Canada Special Contributions, Number 47-1. 1947.

AUB Aubouin, Jean. Geosynclines. New York; Elsevier. 1965.

BAD Badash, Lawrence. Rutherford, Boltwood, and the Age of the Earth. In American Philosophical Society Proceedings, 111: 157-69. 1968.

BAIL Bailey, Edward B. James Hutton -- The Founder of Modern Geology. New York; Elsevier. 1967, c1966.

BAKER Baker, Moses N. and Horton, R. E. Historical Development of Ideas Regarding the Origin of Springs and Spring-water. In Transactions of the American Geophysical Union, Washington, D.C., Part 2: 395-400. 1936.

BAZ El-Baz, Farouk. Gilbert and the Moon. In The Scientific Ideas of G. K. Gilbert, edited by Ellis L. Yochelson. G.S.A. Special Paper 183: 69-80. 1980.

BDCD Segonzac, G. Dunoyerde. Birth and Development of the Concept of Diagenesis (1866-1966). In Earth Science Reviews, 4: 153-201. 1968.

BEER Beer, G. R. de. John Strange, F.R.S., 1732-99. In Royal Society of London Notes and Records, 9: 96-108. 1952.

BEL Beloussov, Vladimir V. Basic Problems in Geotectonics. Translation of the 1954 edition. New York; McGraw Hill. 1962.

BERK Berkland, James O. Elisée Reclus; Neglected Geologic Pioneer and First (?) Continental Drift Advocate. In Geology, 7(4): 189-92. 1979.

BHL Lintner, Stephen F. and Darwin H. Stapleton. Geological Theory and Practice in the Career of Benjamin Henry Latrobe. In Two Hundred Years of Geology in America, Proceedings of the New Hampshire Bicentennial Conference on the History of Geology, edited by Cecil J. Schneer. Hanover, NH; University Press of New England. 1979. pp. 107-119.

BISWAS Biswas, Asit K. History of Hydrology. Amsterdam; North Holland. 1970.

BJG Burke, John G. Origins of the Science of Crystals. Berkeley, CA; University of California Press. 1966.

BLA Black, George W., Jr. Frank Bursley Taylor -- Forgotten Pioneer of Continental Drift. In Journal of Geological Education, 27: 67-70. 1979.

BROME Bromehead, Cyril E. N. Geology in Embryo (up to 1600 A.D.). In Geological Association, London. Proceedings, 55: 89-134. 1945.

BRUSH Brush, Stephen G. Discovery of the Earth's Core. In American Journal of Physics, 48(9): 705-24. 1980.

C&B Chorley, Richard J. and R. P. Beckinsale. G. K. Gilbert's Geomorphology in The Scientific Ideas of G. K. Gilbert, edited by Ellis L. Yochelson. G. S. A. Special Paper 183: 129-42. 1980.

CAM Allen, Don C. The Legend of Noah. In University of Illinois Studies in Language and Literature, 33(3-4). 1963, c1949.

CAR Carozzi, Albert V. De Maillet's Telliamid (1748); The Theory of the Diminution of the Sea. In Endeavour, 29(108): 104-43. 1970.

CAR-2 -----. New Historical Data on the Origin of the Theory of Continental Drift. In Geological Society of America Bulletin, 81: 283-6. 1970.

CBS Bork, Kennard B. Concept of Biotic Succession. In Journal of Geological Education, 32: 213-25. 1984.

CDBN Rupke, Nicolaas A. Continental Drift Before 1900. In Nature, 227: 349-50. 1970.

CDW Yochelson, Ellis L. Charles D. Walcott -- America's Pioneer in Precambrian Paleontology and Stratigraphy. In History of Concepts in Precambrian Geology, edited by W. O. Kupsch and W. A. S. Sarjeant. Geological Association of Canada Special Paper, 19: 261-92. 1979.

CGT North, F. J. Centenary of the Glacial Theory. In Proceedings of the Geological Association, 54: 1-28. 1943.

CGPS LeConte, Joseph. A Century of Geology. In Annual Report of the Smithsonian Institution, 1900: 265-87. 1901 (reprint).

CHA Challinor, John. Uniformitarianism -- the Fundamental Principle of Geology. In International Geological Congress, Report of the Twenty-third Session, Czechoslovakia, 1960. Proceedings, 13: 331-43. 1968.

Chorley, Richard J., Beckinsale, R. P., and Dunn, A. J. The History of the Study of Landforms or the Development of Geomorphology.

CHOR Volume 1. Geomorphology Before Davis. London; Methuen. 1964.

CHOR-2 Volume 2. The Life and Work of William Morris Davis. London; Methuen. 1973.

CHOW Chow, Ven Te. Handbook of Applied Hydrology. New York; McGraw Hill. 1964.

CIR Tomkieff, Sergei I. Classification of Igneous Rocks. In Geological Magazine, 76: 41-8. 1939.

CLA Clark, James M. James Hall of Albany; Geologist and Paleontologist, 1811-1898. Albany, NY; n.p. (privately printed). 1923.

CLOUD Cloud, Preston. Adventures in Earth History. San Francisco, CA; W. H. Freeman. 1970.

CONK Conkin, Barbara M. and James E., eds. Stratigraphy: Foundations and Concepts. New York; Van Nostrand Reinhold Co. 1984.

CORB Corbett, David. Rocks and Men; an Introduction to the History of Geology. Adelaide, Australia; University of Adelaide, Department of Adult Education. 1976.

COX Cox, Leslie R. British Palaeontology: Retrospect and Survey. In Geologists' Association, London. Proceedings, 67: 209-20. 1956-57.

CROSS Cross, Whitman. The Development of Systematic Petrography in the Nineteenth Century. In Journal of Geology, 10: 331-76, 451-9. 1902.

D&B Dott, Robert H. and Batten, Roger L. Evolution of the earth. Second Edition. New York; McGraw Hill. 1976.

DALY Daly, Reginald A. Isostasy Theory in the Making. In Pan-American Geologist, 75(1): 1-7. 1941.

DANA Dana, James D. Manual of Mineralogy. Nineteenth Edition. New York; Wiley. 1977.

DAV Davison, Charles. The Founders of Seismology. Cambridge; The University Press. 1927.

DAVIES Davies, Gordon L. Robert Hooke and His Conception of Earth History. In Geologists´ Association, London. Proceedings, 75: 493-98. 1964.

DEAN Dean, Dennis R. Age of the Earth Controversy. In Annals of Science, 38: 435-56. 1981.

DEAS Deas, Herbert. Crystallography and Crystallographers in England in the Nineteenth Century. In Centaurus, 6: 129-48. 1959.

DEBUS Debus, Allen G. Gabriel Plattes and His Chemical Theory of the Formation of the Earth´s Crust. In Ambix, 9: 162-5. 1961.

DGGS Thams, Johann C., ed. Development of Geodesy and Geophysics in Switzerland. Zurich; Swiss Academy of Natural Science. 1967.

DNS Wilson, John A. Darwinian Natural Selection and Vertebrate Paleontology. In Symposium on the History of American Geology, December 27 and 28, 1958. In Journal of the Washington Academy of Sciences, 49 (7): 231-33. 1959.

DON Donovan, Nowell, and Sanderson, D. Georoots: an Examination of the Origins of Geological Thought in Scotland. Edinburgh; Scottish Office of the Institute of Geological Sciences. 1979.

DRH Dott, R. H., Jr. The Geosyncline -- First Major Geological Concept "Made in America". In Two Hundred Years of Geology in America, Proceedings of the New Hampshire Bicentennial Conference on the History of Geology, edited by Cecil J. Schneer. Hanover, NH; University Press of New England. 1979. pp. 239-64.

DTMG Thwaites, Fredrik T. Development of the Theory of Multiple Glaciation in North America. In Transactions of the Wisconsin Academy of Science, Arts and Letters, 23: 41-164. 1927.

EDEKH White, George W. Early Description and Explanation of Kettle Holes; Charles Whittlesey (1808-1886). In Journal of Glaciology, 5 (37): 119-22. 1964.

EDW Edwards, Wilfred N. The Early History of Palaeontology. London; British Museum. 1967.

EGK Eyles, V. A. The Extent of Geological Knowledge in the Eighteenth Century and the Methods By Which It Was Diffused. In _Toward a History of Geology, Proceedings of the New Hampshire Inter-Disciplinary Conference on the History of Geology, September 7-12, 1967_, edited by Cecil J. Schneer. Cambridge, MA; M.I.T. Press. 1969. pp. 159-83.

EHGT Hansen, Bert. Early History of Glacial Theory in British Geology. In _Journal of Glaciology_, 9 (55): 135-41. 1970.

EOSG Hobbs, William H. Evolution and the Outlook of Seismic Geology. In _American Philosophical Society. Proceedings_, 48: 259-302. 1909.

EPBG-I Challinor, John. The Early Progress of British Geology. I. From Leland to Woodward, 1538-1728. In _Annals of Science_, 9: 124-53. 1953.

EPBG-II -----. The Early Progress of British Geology. II. From Strachey to Michell, 1719-1788. In _Annals of Science_, 10: 1-19. 1954.

EPBG-III -----. The Early Progress of British Geology. III. From Hutton to Playfair, 1788-1802. In _Annals of Science_, 10: 107-48. 1954.

ESSAI Murthy, S. R. N. Earth Science Studies in Ancient India. In _Indian Minerals_, 30: 29-42. 1976.

ETHP Durney, David W. Early Theories and Hypotheses on Pressure-solution Systems. In _Geology_, 6: 369-72. 1978.

EUG Eugster, Hans P. The Beginnings of Experimental Petrology. In _Science_, 173 (3996): 481-9. 1971.

FAIR Fairbridge, Rhodes W., ed. _The Encyclopedia of Geochemistry and Environmental Sciences_. Stroudsburg, PA; Dowden, Hutchison and Ross. 1972.

FCG Glaessner, M. F. and Teichert, C. Geosynclines: A Fundamental Concept in Geology. In _American Journal of Science_, 245: 465-82, 571-91. 1947.

FENT Fenton, Carroll H. and Mildred A. _Giants of Geology_. Garden City, NY; Doubleday and Sons. 1945, 1952.

FGA Hazen, Robert M. The Founding of Geology in America: 1771-1818. _Geological Society of America Bulletin_, 85 (12): 1827-34. 1974.

FOHYAG Merrill, George P. The First One Hundred Years of American Geology. New Haven, CT; Yale University Press. 1924.

FORB Forbes, Robert J. _Studies in Early Petroleum History_. Leiden; E. J. Brill. 1958.

FOREL Forel, F. A. Jean-Pierre Perraudin de Lourtier. In Bulletin de la Société Vaudoise des Sciences Naturelles, 35 (132): 104-13. 1876.

FRANG Frängsmyr, Tore. Geologi och Skapelsetro. (Geology and the Doctrine of Creation). Stockholm; Almqvist and Wiksell. 1969.

FRANG-2 -----. Swedish Science in the Eighteenth Century. In History of Science, 12: 29-42. 1974.

FRMS Albury, W. R. and Oldroyd, D.R. From Renaissance Mineral Studies to Historical Geology in the Light of Michel Foucault's "The Order of Things". In British Journal for the History of Science, 10: 187-215. 1977.

FYPPP Bascom, Florence. Fifty Years of Progress in Petrography and Petrology, 1876-1926. In Johns Hopkins University Studies in Geology, 8: 33-82. 1927.

GEI Geikie, Archibald. Founders of Geology. Second Edition. London; Macmillan. 1905.

GEI L&P -----. Lamarck and Playfair. In Geological Magazine, 43: 145-53, 193-202. 1906.

GER Gerstner, Patsy A. A Dynamic Theory of Mountain Building: Henry Darwin Rogers, 1842. In Isis, 66: 26-37. 1975.

GHT Porter, Roy S. George Hoggart Toulmin's Theory of Man and the Earth in the Light of the Development of British Geology. In Annals of Science, 35 (4): 339-52. 1978.

GIA Carozzi, Albert V. Glaciology and the Ice Age. In Journal of Geological Education, 32: 158-70. 1984.

GKES Barrell, Joseph. A Century of Geology; the Growth of Knowledge of Earth Structure. In American Journal of Science, 196: 133-70. 1918.

GKG Pyne, Stephen J. Certain Allied Problems in Mechanics: Grove Karl Gilbert at the Henry Mountains. In Two Hundred Years of Geology in America, Proceedings of the New Hampshire Bicentennial Conference on the History of Geology, edited by Cecil J. Schneer. Hanover, NH; University Press of New England. 1979. pp. 225-28.

GLD Davies, Gordon L. Early British Geomorphology, 1578-1705. In Geographical Journal, 132: 252-62. 1966.

GNC Greene, Mott T. Geology in the Nineteenth Century; Changing Views of a Changing World. Ithaca, NY; Cornell University Press. 1982.

GORT Gortani, Michele. Italian Pioneers in Geology and Mineralogy. In Journal of World History, 7: 503-19. 1963.

GRA Graham, R. P. D. The Development of Mineralogical Science. In Proceedings and Transactions of the Royal Society of Canada, Volume 28, Third Series, Section 4: 33-42. 1934.

GREG Gregory, Herbert E. A Century of Geology -- Steps of Progress in the Interpretation of Land Forms. In American Journal of Science, 196: 104-32. 1918.

HAL Hallam, Anthony. A Revolution in the Earth Sciences from Continental Drift to Plate Tectonics. Oxford; Clarendon Press. 1973.

HAN Hanna, G. Dallas. Early Reference to the Theory That Diatoms Are the Source of Bituminous Substances. In American Association of Petroleum Geologists. Bulletin, 12 (5): 555-56. 1928.

HARR Harrington, John W. The First, First Principles of Geology. In American Journal of Science, 265: 449-61. 1967.

HAUS Hausen, Hans. The History of Geology and Mineralogy in Finland 1828-1918. Helsinki; Societas Scientiarum Fennica. 1968.

HAZ Hazen, Robert M. North American Geology, Early Writings. Stroudsburg, PA; Dowden, Hutchinson & Ross. 1979.

HCM Dietrich, R. V. History of Concepts: Migmatites. In History of Concepts in Precambrian Geology, edited by W.O. Kupsch and W.A.S. Sarjeant. Geological Association of Canada Special Paper, 19: 51-63. 1979.

HCMP Manten, A. A. Half a Century of Modern Palynology. In Earth-Science Reviews, 2: 277-316. 1966.

HDMF Anderson, Jay E., Jr. History of the Development of the Metamorphism Facies Concept. In Compass, 38(3): 149-56. 1961.

HED Hedberg, Hollis D. Influence of Torbern Bergman (1735-1784) on Stratigraphy. In Stockholm Contributions in Geology, 20: 19-47. 1969a.

HER Fairchild, Herman Le Roy. The Geological Society of America, 1888-1930. New York; Geological Society of America. 1932.

HES Hall, D. H. History of the Earth Sciences During the Scientific and Industrial Revolutions with Special Emphasis on the Physical Geosciences. New York; Elsevier. 1976.

HIGG Higgins, G. E. and Saunders, J.B. Mud Volcanoes -- Their Nature and Origin. In Naturforschende Gesellschaft Basel. Verhandlungen, 84 (1): 101-52. 1974.

HLF Fairchild, Herman L. Glacial Geology in America. In American Geologist, 22: 154-89. 1898.

HMPD Marsh, O. C. History and Methods of Palaeontological Discovery. In American Journal of Science, Series 3, 18: 323-59. 1879.

HOOY Hooykaas, Reijer. Catastrophism in Geology, Its Scientific Character in Relation to Actualism and Uniformitarianism. Koninklijke Nederlandse Akademie van Wetenschappen, Afdeling Letterkunde, Mededelingen, 33 (7): 271-316. 1970.

HSP Tomkieff, Sergei I. A Historical Survey of Petrology. 1936 edition translated by Frants Levinson-Lessing. Edinburgh; Oliver and Boyd. 1954.

HTOD Crook, Thomas. History of the Theory of Ore Deposits. London; T. Murby & Co. 1933.

HUBB Hubbert, M. King. Critique of the Principle of Uniformity. In Philosophy of Geohistory: 1785-1970, edited by Claude C. Albritton Jr. Stroudsburg, PA; Dowden, Hutchinson and Ross, Inc. 1975. pp 225-55.

HUNT Hunt, Charles B. G. K. Gilbert, on Laccoliths and Intrusive Structures. In The Scientific Ideas of G. K. Gilbert, edited by Ellis L. Yochelson. G.S.A. Special Paper 183: 25-34. 1980.

HURL Hurlbut, Cornelius S. and Switzer, George S. Gemology. New York; Wiley. 1979.

JJGO White, George W. John Josselyn's Geological Observations. In Illinois Academy of Science Transactions, 48: 173-82. 1954.

JONES Jones, F. Wood. John Hunter as a Geologist. In Annals of the Royal College of Surgeons, 12: 219-44. 1953.

JORD Jordan, William M. Geology and the Industrial-Transportation Revolution in Early to Mid Nineteenth Century Pennsylvania. In Two Hundred Years of Geology in America, Proceedings of the New Hampshire Bicentennial Conference on the History of Geology, edited by Cecil J. Schneer. Hanover, NH; University Press of New England. 1979. pp. 91-103.

JUDD Judd, John W. Henry Clifton Sorby and the Birth of Microscopical Petrology. In Geological Magazine, 5: 193-204. 1908.

KAR Karunakaran, C. and Murthy, S. R. N. On Some Earth Science Observations in Sanskrit Texts. In Records of the Geological Survey of India, 109 (2): 82-103. 1977.

KRY Krynine, Paul D. On the Antiquity of Sedimentation and Hydrology. In Geological Society of America Bulletin, 71: 1721-6. 1960.

KUNZ Kunz, George F. Life and Work of Haüy. In American Mineralogist, 3: 61-89. 1918.

LBS Hunt, Charles B. G. K. Gilbert's Lake Bonneville Studies. In The Scientific Ideas of G. K. Gilbert, edited by Ellis L. Yochelson. G.S.A. Special Paper 183: 45-59. 1980.

LECT White, George W. Lewis Evans' Contributions to Early American Geology, 1743-1755. In Illinois Academy of Science Transactions, 44: 152-58. 1951.

LEEA -----. Lewis Evans' Early American Notice of Isostasy. In Science, 114 (2960): 302-3. 1951.

LIND Lindroth, S. Urban Hiärne 1641-1724. In Swedish Men of Science: 42-9. Stockholm; Swedish Institute. 1952.

MAB Mabey, Don R. Pioneering Work of G. K. Gilbert on Gravity and Isostasy. In The Scientific Ideas of G. K. Gilbert, edited by Ellis L. Yochelson. G.S.A. Special Paper 183: 61-68. 1980.

MARG Margerie, Emanuel de. Three Stages in the Evolution of Alpine Geology. In Quarterly Journal of the Geological Society of London, 102: xcvii-cxiv. 1946.

MATH Mather, Kirtley and Mason, Shirley. A Source Book in Geology. New York; McGraw Hill. 1939.

MBC White, George W. William McClure's Concept of Primitive Rocks ("Basement Concept"). In History of Concepts in Precambrian Geology, edited by W. O. Kupsch and W. A. S. Sarjeant. Geological Association of Canada Special Paper, 19: 251-9. 1979.

MCR Oldroyd, D. R. Mineralogy and the 'Chemical Revolution'. In Centaurus, 19: 54-71. 1975.

MERR Merrill, George P. Contributions to the History of American Geology. Washington, D.C.; Government Printing Office. 1906.

MEYER Meyerhoff, A. A. Arthur Holmes: Originator of Spreading Ocean Floor Hypothesis. In Journal of Geophysical Research, 73 (20): 6563-5. 1968.

MHG LaRocque, Aurèle. Milestones in the History of Geology; a Chronologic List of Important Events in the Development of Geology. In Journal of Geological Education, 22 (5): 195-203. 1974.

MILL Miller, Samuel. Brief Retrospect of the Eighteenth Century. New York; T. & J. Swords. 1803.

MOR Morrell, J.B. Reflections on the History of Scottish Science. In History of Science, 12: 81-94. 1974.

MORT Morton, W. L. Henry Youle Hind (1823-1908). In Geological Association of Canada Proceedings, 23: 25-29. 1971.

MSSC Oldroyd, D. R. Some NeoPlatonic and Stoic Influences on Mineralogy in the Sixteenth and Seventeenth Centuries. In Ambix, 21 (2-3): 128-56. 1974.

MUR Murty, K. S. History of Geoscience Information in India. In Geoscience Information: A State of the Art Review, edited by A. P. Harvey and J. A. Diment. London; Broad Oak Press. 1979. pp. 51-60.

NCDAFG Brush, Stephen G. Nineteenth Century Debates about the Inside of the Earth. In Annals of Science, 36(3): 225-54. 1979.

NEILL Neill, Patrick. Biographical Account of Mr. Williams, the Mineralogist. In Annals of Philosophy, 4: 81-3. 1814.

NEVE Neve, Michael and Porter, R. S. Alexander Catcott: Glory and Geology. In British Journal for the History of Science, 10: 47-70. 1877.

OHER Ohern, D. W. Fifty Years of Petroleum Geology. In Johns Hopkins University Studies in Geology, 8: 83-120. 1927.

OLD NS Oldroyd, D. R. A Note on the Status of A. F. Cronstedt´s Simple Earths and His Analytical Methods. In Isis, 65: 506-12. 1974.

OLD RHG -----. Historicism and the Rise of Historical Geology, Parts I and II. In History of Science, 17 (3): 191-213 and 17 (4): 227-257. 1979.

OPG Brown, Bahngrell W. Origin of Petroleum Geology. In Bulletin of the American Association of Petroleum Geologists, 56(3): 566-68. 1972.

OSPO Ospovat, Alexander M. Werner´s Concept of the Basement Complex. In History of Concepts in Precambrian Geology, edited by W. O. Kupsch and W. A. S. Sarjeant. Geological Survey of Canada Special Paper, 19: 161-70. 1979.

OSR Adams, Frank D. The Origin of Springs and Rivers. In Fennia, 50(1): 1-18. 1928.

OWEN Owen, Edgar W. Trek of the Oil Finders; a History of Exploration for Petroleum. Tulsa, OK; American Association of Petroleum Geologists. 1975.

P&S Press, Frank and Sever, Raymond. Earth. San Francisco, CA; W. H. Freeman. 1974.

PBG Challinor, John. Progress of British Geology During the Early Part of the Nineteenth Century. In Annals of Science, 26: 177-234. 1970.

PHIL Phillips, Frank C. An Introduction to Crystallography. Third Edition. New York; Wiley. 1963.

POGS Miller, Benjamin Leroy. Progress in Ore Genesis Studies. In Johns Hopkins University Studies in Geology, 8: 121-35. 1927.

PORT Porter, Roy S. The Making of Geology; Earth Science in Britain, 1660-1815. Cambridge; Cambridge University Press. 1977.

PORT-2 -----. William Hobbs of Weymouth and His "The Earth Generated and Anatomized". In Society for the Bibliography of Natural History Journal, 7: 333-41. 1976.

PPHG Regnell, Gerhard. On the Position of Palaeontology and Historical Geology in Sweden. In Arkiv för Mineralogi och Geologi, 1 (1): 1-64. 1947.

PRG Harrington, John W. The PreNatal Roots of Geology; a Study in the History of Ideas. In American Journal of Science, 267 (5): 592-7. 1969.

PSG Matthews, Edward B. Progress in Structural Geology. In Johns Hopkins University Studies in Geology, 8: 137-61. 1927.

PYNE Pyne, Stephen J. From the Grand Canyon to the Marianas Trench: the Earth Sciences After Darwin. In The Sciences in the American Context: New Perspectives, edited by Nathan Reingold. Washington, D.C.; Smithsonian Institution Press. 1979. pp. 165-192.

RAM-1 Ramsay, Andrew C. Passages in the History of Geology: Being an Inaugural Lecture, University College, London, 1848. London; Printed for Taylor and Walton. 1848.

RAM-2 -----. Passages in the History of Geology: Being an Introductory Lecture at University College, London, in Continuation of the Inaugural Lecture of 1848. London; Printed for Taylor, Walton and Maberly. 1849.

RAN Ranalli, Georgio. Robert Hooke and the Huttonian Theory. In Journal of Geology, 90 (3): 319-23. 1982.

REG Regnell, Gerhard. Primaeval Earth as Conceived by 18th Century Naturalists. In History of Concepts in Precambrian Geology, edited by W. O. Kupsch and W.A.S. Sarjeant, Geological Association of Canada Special Paper 19: 171-80. 1979.

RGIF Thomas, H. Hamshaw. Rise of Geology and Its Influence on Contemporary Thought. In *Annals of Science*, 5: 325-41. 1947.

RHAP Owen, Edgar W. Remarks on the History of American Petroleum Geology. In *Journal of the Washington Academy of Science*, 49 (7): 256-60. 1959.

RHRE Carozzi, Albert V. Robert Hooke, Rudolf Eric Raspe, and the Concept of "Earthquakes". In *Isis*, 61(1): 85-91. 1970.

RISE Schneer, Cecil J. Rise of Historical Geology. In *Isis*, 45: 256-67. 1954.

RJA Sweet, Jessie M. and Waterston, C. Robert Jameson´s Approach to the Wernerian Theory of the Earth, 1796. In *Annals of Science*, 23: 81-95. 1967.

ROD Rodolico, Francesco. Niels Stensen, Founder of the Geology of Tuscany. In *Dissertations on Steno as Geologist*. In *Acta Historica Scientiarium Naturalium et Medicinalium*, 23: 237-43. 1971.

ROWL Rowlinson, J. S. Theory of Glaciers. In *Royal Society of London Notes and Record*, 26 (2): 189-204. 1971.

RPM Prior, G. T. Review of Progress of Mineralogy from 1864 to 1918. In *Geological Magazine*, 56 (1): 10-18. 1919.

RPS Pirsson, Louis V. Rise of Petrology as a Science. In *American Journal of Science*, 46: 222-39. 1918.

RTP Skinner, Herbert C. Raymond Thomassy and the Practical Geology of Louisiana. In *Two Hundred Years of Geology in America, Proceedings of the New Hampshire Bicentennial Conference on the History of Geology*, edited by Cecil J. Schneer. Hanover, NH; University Press of New England. 1979. pp. 201-11.

RUD Rudwick, Martin J. S. *The Meaning of Fossils*. Second Edition. New York; Neale Watson. 1976.

SARJ Sarjeant, William A. S. and Harvey, A. P. Uriconian and Langmyndian: A History of the Study of the Precambrian Rocks of the Welsh Borderland. In *History of Concepts in Precambrian Geology*, edited by W. O. Kupsch and W. A. S. Sarjeant. *Geological Association of Canada Special Paper*, 19: 181-224. 1979.

SCH Schneer, Cecil J. Introduction. In *Toward a History of Geology, Proceedings of the New Hampshire Inter-Disciplinary Conference on the History of Geology, September 7-12, 1967*, edited by Cecil J. Schneer. Cambridge, MA; M.I.T. Press. 1969. pp. 1-18.

CHERZ Scherz, Gustav. Niels Stensens Riesen. In Dissertations on Steno as Geologist. In Acta Historica Scientiarum Naturalium et Medicinalium, 23: 9-137. 1971.

CHU Schuchert, Charles. A Century of Geology; the Progress of Historical Geology in North America. In American Journal of Science, 196: 45-103. 1918.

EDD Seddon, George. Abraham Gottlob Werner; History and Folk-history. In Geological Society of Australia Journal, 20 (4): 381-95. 1973.

IEG Siegfried, Robert and R. H. Dott, Jr. Humphry Davy as Geologist, 1805-29. In British Journal for the History of Science, 9: 219-27. 1976.

IMP Simpson, George G. Uniformitarianism; an Inquiry into Principle, Theory and Method in Geohistory and Biohistory. In Philosophy of Geohistory: 1785-1970, edited by Claude C. Albritton, Jr. Stroudsburg, PA: Dowden, Hutchinson and Ross, Inc. 1975.

LEEP Sleep, Mark C. W. Sir William Hamilton (1730-1803), His Work and Influence in Geology. In Annals of Science, 25: 319-38. 1969.

LPG Rudwick, Martin J. S. The Strategy of Lyell's Principles of Geology. In Isis, 61 (1): 5-33. 1970.

MAJ Greene, John C. and Burke, John G. The Science of Minerals in the Age of Jefferson. In American Philosophical Society Transactions, 68 (4): 1-113. 1978.

ND Jahn, Melvin E. Some Notes on Dr. Scheuchzer and on Homo diluvii testis. In Toward a History of Geology, Proceedings of the New Hampshire Inter-Disciplinary Conference on the History of Geology, September 7-12, 1967, edited by Cecil J. Schneer. Cambridge, MA; M.I.T. Press. 1969. pp. 193-213.

OC Burstyn, Harold L. and Susan B. Schlee. The Study of Ocean Currents in America Before 1930. In Two Hundred Years of Geology in America, Proceedings of the New Hampshire Bicentennial Conference on the History of Geology, edited by Cecil J. Schneer. Hanover, NH: University Press of New England. 1979. pp. 115-55.

PMS Oldroyd, D. R. Some Phlogistic Mineralogical Schemes Illustrative of the Concept of 'Earth' in the Seventeenth and Eighteenth Centuries. In Annals of Science, 30: 269-306. 1974.

STN Sarton, George. Introduction to the History of Science. Volumes 1-3. Baltimore, MD: Published for the Carnegie Institution of Washington by Williams and Wilkins Co. 1927.

STODD Stoddart, D. R. Darwin, Lyell, and the Geological Significance of Coral Reefs. In British Journal for the History of Science, 9 (2): 199-218. 1976.

STOKES Stokes, Evelyn. The Six Days and The Deluge; Some Ideas on Earth History in the Royal Society of London 1660-1775. In Earth Science Journal, 3 (1): 13-39. 1969.

STP Debus, Allen G. Edward Jorden and the Fermentation of Metals: an Iatrochemical Study of Terrestrial Phenomena. In Toward a History of Geology, Proceedings of the New Hampshire Inter-Disciplinary Conference on the History of Geology, September 7-12, 1967, edited by Cecil J. Schneer. Cambridge, MA; M.I.T. Press. 1969. pp. 100-21

STRO Stroh, A. H., editor. Emmanuel Swedenborg as a Scientist. Stockholm; Aftonbladets Trycheri. 1908-11.

TAY Taylor, John. Report on the State of Knowledge Respecting Mineral Veins. In Report of the British Association, Third Meeting, 1833: 1-25. 1834.

TBS Hedberg, Hollis D. The Influence of Torbern Bergman (1735-1784) on Stratigraphy: a Résumé. In Toward a History of Geology, Proceedings of the New Hampshire Inter-Disciplinary Conference on the History of Geology, September 7-12, 1967, edited by Cecil J. Schneer. Cambridge, MA; M.I.T. Press. 1969. pp. 186-92.

TDG Tikhomirov, V. V. The Development of the Geological Sciences in the U. S. S. R. from Ancient Times to the Middle of the Nineteenth Century. In Toward a History of Geology, Proceedings of the New Hampshire Inter-Disciplinary Conference on the History of Geology, September 7-12, 1967, edited by Cecil J. Schneer. Cambridge, MA; M.I.T. Press. 1969. pp. 357-85.

TER Kelly, Sister Suzanne. Theories of the Earth in Renaissance Cosmologies. In Toward a History of Geology, Proceedings of the New Hampshire Inter-Disciplinary Conference on the History of Geology, September 7-12, 1967, edited by Cecil J. Schneer. Cambridge, MA; M.I.T. Press. 1969. pp. 214-25.

TIK Tikhomirov, V. V. Development of Geological Sciences During the First Half of the Nineteenth Century. In International Geological Congress, Report of the Twenty-third Session, Czechoslovakia, Section 13. Proceedings: 319-29. 1968.

TND Taylor, Kenneth L. Nicolas Desmarest and Geology in the Eighteenth Century. In Toward a History of Geology, Proceedings of the New Hampshire Inter-Disciplinary Conference on the History of Geology, September 7-12, 1967, edited by Cecil J. Schneer. Cambridge, MA; M.I.T. Press. 1969. pp.339-56.

TODH Emmons, Samuel F. Theories of Ore Deposition Historically Considered. In Annual Report of the Smithsonian Institute for 1904: 309-36. 1905.

JHS Tomkieff, Sergei I. Unconformity -- An Historical Study. In Geologists´ Association, London. Proceedings. 73 (4): 383-417. 1962.

JPG Alexandrov, Eugene A. 100th Anniversary of Observations on Petroleum Geology in the U. S. A. by Dimitriy I. Mendeleyev. In Two Hundred Years of Geology in America, Proceedings of the New Hampshire Bicentennial Conference on the History of Geology, edited by Cecil J. Schneer. Hanover, NH: University Press of New England. 1979. pp. 285-88.

VAID Shastri, Vaidyanath. Sciences in the Vedas. New Delhi; Sarvadeshik Arya Pratinidhi Sabha. 1970.

VON Baumgärtel, Hans. Alexander von Humboldt: Remarks on the Meaning of Hypothesis in His Geological Researches. In Toward a History of Geology, Proceedings of the New Hampshire Inter-Disciplinary Conference on the History of Geology, September 7-12, 1967, edited by Cecil J. Schneer. Cambridge, MA; M.I.T. Press. 1969. pp. 19-35.

WALL Wallace, Robert E. G. K. Gilbert´s Studies of Faults, Scarps, and Earthquakes. In The Scientific Ideas of G. K. Gilbert, edited by Ellis L. Yochelson. G. S. A. Special Paper 183: 35-44. 1980.

WATER Waterston, Charles D. William Thomson (1761-1806) a Forgotten Benefactor. In University of Edinburgh Journal, 22 (2): 122-34. 1965.

WDC Page, Leroy E. Diluvialism and Its Critics in Great Britain in the Early Nineteenth Century. In Toward a History of Geology, Proceedings of the New Hampshire Inter-Disciplinary Conference on the History of Geology, September 7-12, 1967, edited by Cecil J. Schneer. Cambridge, MA; M.I.T. Press. 1969. pp. 257-71.

WEG Wegmann, Eugene. Changing Ideas about Moving Shorelines. In Toward a History of Geology, Proceedings of the New Hampshire Inter-Disciplinary Conference on the History of Geology, September 7-12, 1967, edited by Cecil J. Schneer. Cambridge, MA; M.I.T. Press. 1969. pp. 386-414.

WKK Ospovat, Alexander M. Reflections on A. G. Werner´s "Kurze Klassifikation". In Toward a History of Geology, Proceedings of the New Hampshire Inter-Disciplinary Conference on the History of Geology, September 7-12, 1967, edited by Cecil J. Schneer. Cambridge, MA; M.I.T. Press. 1969. pp. 242-56.

WMC Burke, John G. Mineral Classification in the Early Nineteenth Century. In Toward a History of Geology, Proceedings of the New Hampshire Inter-Disciplinary Conference on the History of Geology, September 7-12, 1967, edited by Cecil J. Schneer. Cambridge, MA; M.I.T. Press. 1969. pp. 62-77.

WOOD Woodward, Horace B. History of Geology. New York; G. P. Putnam's Sons. 1911.

WOOD-2 Wood, Robert M. Coming Apart at the Seams (Alfred Wegener, the Founder of Continental Drift -- Or Was He?). In New Scientist 85(1191): 252-54. 1980.

WRB Brice, William R. Bishop Ussher, John Lightfoot and the Age of Creation. In Journal of Geological Education, 30(1): 18-24. 1982.

WSSB Davies, A. Morley. The Geological Life-work of Sydney Savory Buckman. In Geologists' Association, London. Proceedings, 41 221-40. 1930.

YANS Yanshin, A. L., Garetskiy, R. G., and Shiezinger, A. Y. Role of the U. S. S. R. Academy of Sciences in Development of the Theory of Platforms, and Some Present Aspects of This Theory. In Geotectonics, 3: 127-36. 1974.

YODER Yoder, Hatten S. Jr. Experimental Mineralogy: Achievements and Prospects. In Bulletin de Minéralogie, 103 (1): 5-26. 1980.

ZIT Zittel, Karl A. von. History of Geology and Palaeontology to the End of the Nineteenth Century. London; Walter Scott. 1901.

BIBLIOGRAPHY OF SOURCES CITED

Adams, Frank D. The Birth and Development of the Geological Sciences. New York: Dover. 1954, c1938.

-----. The Origin of Springs and Rivers. In Fennia, 50(1): 1-18. 1928.

Agnew, Allen F. G.K. Gilbert and Groundwater, or ´I have drawn this map with great reluctance.´ In The Scientific Ideas of G. K. Gilbert, edited by Ellis Y. Yochelson. G. S. A. Special Paper 183: 81-91. 1980.

Albritton, Claude C. The Abyss of Time. San Francisco, CA: Freeman, Cooper. 1980.

-----. The Fabric of Geology, Reading, MA; Addison-Wesley, Inc. 1963.

-----. Philosophy of Geohistory: 1785-1970. Stroudsberg, PA; Dowden, Hutchinson & Ross, Inc. 1975.

Albury, W. R. and Oldroyd, D.R. From Renaissance Mineral Studies to Historical Geology in the Light of Michel Foucault´s "The Order of Things". In British Journal for the History of Science, 10: 187-215. 1977.

Alcock, Frederick J. A Century in the History of the Geological Survey of Canada. In National Museum of Canada Special Contributions, Number 47-1. 1947.

Alexandrov, Eugene A. 100th Anniversary of Observations on Petroleum Geology in the U. S. A. by Dimitriy I. Mendeleyev. In Two Hundred Years of Geology in America, Proceedings of the New Hampshire Bicentennial Conference on the History of Geology, edited by Cecil J. Schneer. Hanover, NH: University Press of New England. 1979. pp. 285-88.

Allen, Don C. The Legend of Noah. In University of Illinois Studies in Language and Literature, 33(3-4). 1963, c1949.

Anderson, Jay E., Jr. History of the Development of the Metamorphism Facies Concept. In Compass, 38(3): 149-56. 1961.

Aubouin, Jean. Geosynclines. New York; Elsevier. 1965.

Badash, Lawrence. Rutherford, Boltwood, and the Age of the Earth. In American Philosophical Society Proceedings, 111: 157-69. 1968.

Bailey, Edward B. James Hutton -- The Founder of Modern Geology. New York; Elsevier. 1967, c1966.

Baker, Moses N. and Horton, R. E. Historical Development of Ideas Regarding the Origin of Springs and Spring-water. In Transactions of the American Geophysical Union, Washington, D.C., Part 2: 395-400. 1936.

Barrell, Joseph. A Century of Geology; the Growth of Knowledge of Earth Structure. In American Journal of Science, 196: 133-70. 1918.

Bascom, Florence. Fifty Years of Progress in Petrography and Petrology, 1876-1926. In Johns Hopkins University Studies in Geology, 8: 33-82. 1927.

Baumgärtel, Hans. Alexander von Humboldt: Remarks on the Meaning of Hypothesis in His Geological Researches. In Toward a History of Geology, Proceedings of the New Hampshire Inter-Disciplinary Conference on the History of Geology, September 7-12, 1967, edited by Cecil J. Schneer. Cambridge, MA; M.I.T. Press. 1969. pp. 19-35.

Beer, G. R. de. John Strange, F.R.S., 1732-99. In Royal Society of London Notes and Records, 9: 96-108. 1952.

Beloussov, Vladimir V. Basic Problems in Geotectonics. Translation of the 1954 edition. New York; McGraw Hill. 1962.

Berkland, James O. Elisée Reclus; Neglected Geologic Pioneer and First (?) Continental Drift Advocate. In Geology, 7(4): 189-92. 1979.

Biswas, Asit K. History of Hydrology. Amsterdam; North Holland. 1970.

Black, George W., Jr. Frank Bursley Taylor -- Forgotten Pioneer of Continental Drift. In Journal of Geological Education, 27: 67-70. 1979.

Bork, Kennard B. Concept of Biotic Succession. In Journal of Geological Education, 32: 213-25. 1984.

Brice, William R. Bishop Ussher, John Lightfoot and the Age of Creation. In Journal of Geological Education, 30(1): 18-24. 1982.

Bromehead, Cyril E. N. Geology in Embryo (up to 1600 A.D.). In Geological Association, London. Proceedings, 55: 89-134. 1945.

Brown, Bahngrell W. Origin of Petroleum Geology. In Bulletin of the American Association of Petroleum Geologists, 56(3): 566-68. 1972.

Brush, Stephen G. Discovery of the Earth's Core. In American Journal of Physics, 48(9): 705-24. 1980.

-----. Nineteenth Century Debates about the Inside of the Earth. In Annals of Science, 36(3): 225-54. 1979.

Burke, John G. Mineral Classification in the Early Nineteenth Century. In Toward a History of Geology, Proceedings of the New Hampshire Inter-Disciplinary Conference on the History of Geology, September 7-12, 1967, edited by Cecil J. Schneer. Cambridge, MA; M.I.T. Press. 1969. pp. 62-77.

-----. Origins of the Science of Crystals. Berkeley, CA; University of California Press. 1966.

Burstyn, Harold L. and Susan B. Schlee. The Study of Ocean Currents in America Before 1930. In Two Hundred Years of Geology in America, Proceedings of the New Hampshire Bicentennial Conference on the History of Geology, edited by Cecil J. Schneer. Hanover, NH: University Press of New England. 1979. pp. 115-55.

Carozzi, Albert V. Agassiz´s Influence on Geological Thinking in the Americas. Archives des Sciences, 27(1): 5-38. 1974.

-----. De Maillet´s Telliamid (1748); The Theory of the Diminution of the Sea. In Endeavour, 29(108): 104-43.

-----. Glaciology and the Ice Age. In Journal of Geological Education, 32: 158-70. 1984.

-----. New Historical Data on the Origin of the Theory of Continental Drift. In Geological Society of America Bulletin, 81: 283-6. 1970.

-----. Robert Hooke, Rudolf Eric Raspe, and the Concept of "Earthquakes". In Isis, 61(1): 85-91. 1970.

Challinor, John. The Early Progress of British Geology. I. From Leland to Woodward, 1538-1728. In Annals of Science, 9: 124-53. 1953.

-----. The Early Progress of British Geology. II. From Strachey to Michell, 1719-1788. In Annals of Science, 10: 1-19. 1954.

-----. The Early Progress of British Geology. III. From Hutton to Playfair, 1788-1802. In Annals of Science, 10: 107-48. 1954.

-----. Progress of British Geology During the Early Part of the Nineteenth Century. In Annals of Science, 26: 177-234. 1970.

-----. Uniformitarianism -- the Fundamental Principle of Geology. In International Geological Congress, Report of the Twenty-third Session, Czechoslovakia, 1960. Proceedings, 13: 331-43. 1968.

Chorley, Richard J. and R. P. Beckinsale. G. K. Gilbert´s Geomorphology in The Scientific Ideas of G. K. Gilbert, edited by Ellis L. Yochelson. G. S. A. Special Paper 183: 129-42. 1980.

Chorley, Richard J., Beckinsale, R. P., and Dunn, A. J. The History of the Study of Landforms or the Development of Geomorphology. Volumes 1-2. London; Methuen. 1964, 1973.

Chow, Ven Te. Handbook of Applied Hydrology. New York; McGraw Hill. 1964.

Clark, James M. James Hall of Albany; Geologist and Paleontologist, 1811-1898. Albany, NY; n.p. (privately printed). 1923.

Cloud, Preston. Adventures in Earth History. San Francisco, CA; W. H. Freeman. 1970.

Conkin, Barbara M. and James E., eds. Stratigraphy: Foundations and Concepts. New York; Van Nostrand Reinhold Co. 1984.

Corbett, David. Rocks and Men; an Introduction to the History of Geology. Adelaide, Australia; University of Adelaide, Department of Adult Education. 1976.

Cox, Leslie R. British Palaeontology: Retrospect and Survey. In Geologists' Association, London. Proceedings, edited by W. O. Kupsch and W. A. S. Sarjeant, 67: 209-20. 1956-57.

Crook, Thomas. History of the Theory of Ore Deposits. London; T. Murby & Co. 1933.

Cross, Whitman. The Development of Systematic Petrography in the Nineteenth Century. In Journal of Geology, 10: 331-76, 451-9. 1902.

Daly, Reginald A. Isostasy Theory in the Making. In Pan-American Geologist, 75(1): 1-7. 1941.

Dana, James D. Manual of Mineralogy. Nineteenth Edition. New York; Wiley. 1977.

Davies, A. Morley. The Geological Life-work of Sydney Savory Buckman. In Geologists' Association, London. Proceedings, 41: 221-40. 1930.

Davies, Gordon L. Early British Geomorphology, 1578-1705. In Geographical Journal, 132: 252-62. 1966.

-----. Robert Hooke and His Conception of Earth History. In Geologists' Association, London. Proceedings, 75: 493-98. 1964.

Davison, Charles. The Founders of Seismology. Cambridge; The University Press. 1927.

Dean, Dennis R. Age of the Earth Controversy. In Annals of Science, 38: 435-56. 1981.

Deas, Herbert. Crystallography and Crystallographers in England in the Nineteenth Century. In Centaurus, 6: 129-48. 1959.

Debus, Allen G. Edward Jorden and the Fermentation of Metals: an Iatrochemical Study of Terrestrial Phenomena. In Toward a History of Geology, Proceedings of the New Hampshire Inter-Disciplinary Conference on the History of Geology, September 7-12, 1967, edited by Cecil J. Schneer. Cambridge, MA; M.I.T. Press. 1969. pp. 100-21

-----. Gabriel Plattes and His Chemical Theory of the Formation of the Earth's Crust. In Ambix, 9: 162-5. 1961.

Dietrich, R. V. History of Concepts: Migmatites. In History of Concepts in Precambrian Geology, edited by W.O. Kupsch and W.A.S. Sarjeant. Geological Association of Canada Special Paper, 19: 51-63. 1979.

Donovan, Nowell, and Sanderson, D. Georoots: an Examination of the Origins of Geological Thought in Scotland. Edinburgh; Scottish Office of the Institute of Geological Sciences. 1979.

Dott, Robert H. and Batten, Roger L. Evolution of the earth. Second Edition. New York; McGraw Hill. 1976.

Dott, R. H., Jr. The Geosyncline -- First Major Geological Concept "Made in America". In Two Hundred Years of Geology in America, Proceedings of the New Hampshire Bicentennial Conference on the History of Geology, edited by Cecil J. Schneer. Hanover, NH; University Press of New England. 1979. pp. 239-64.

Durney, David W. Early Theories and Hypotheses on Pressure-solution Systems. In Geology, 6: 369-72. 1978.

Edwards, Wilfred N. The Early History of Palaeontology. London; British Museum. 1967.

El-Baz, Farouk. Gilbert and the Moon. In The Scientific Ideas of G. K. Gilbert, edited by Ellis L. Yochelson. G.S.A. Special Paper 183: 69-80. 1980.

Emmons, Samuel F. Theories of Ore Deposition Historically Considered. In Annual Report of the Smithsonian Institute for 1904: 309-36. 1905.

Eugster, Hans P. The Beginnings of Experimental Petrology. In Science, 173 (3996): 481-9. 1971.

Eyles, V. A. The Extent of Geological Knowledge in the Eighteenth Century and the Methods By Which It Was Diffused. In Toward a History of Geology, Proceedings of the New Hampshire Inter-Disciplinary Conference on the History of Geology, September 7-12, 1967, edited by Cecil J. Schneer. Cambridge, MA; M.I.T. Press. 1969. pp. 159-83.

Fairbridge, Rhodes W., ed. The Encyclopedia of Geochemistry and Environmental Sciences. Stroudsburg, PA; Dowden, Hutchison and Ross. 1972.

Fairchild, Herman L. The Geological Society of America, 1888-1930. New York; Geological Society of America. 1932.

-----. Glacial Geology in America. In American Geologist, 22: 154-89. 1898.

Fenton, Carroll H. and Mildred A. Giants of Geology. Garden City, NY; Doubleday and Sons. 1945, 1952.

Forbes, Robert J. Studies in Early Petroleum History. Leiden; E. J. Brill. 1958.

Forel, F. A. Jean-Pierre Perraudin de Lourtier. In Bulletin de la Société Vaudoise des Sciences Naturelles, 35 (132): 104-13. 1876.

Frängsmyr, Tore. Geologi och Skapelsetro. (Geology and the Doctrine of Creation). Stockholm; Almqvist and Wiksell. 1969.

-----. Swedish Science in the Eighteenth Century. In History of Science, 12: 29-42. 1974.

Geikie, Archibald. Founders of Geology. Second Edition. London; Macmillan. 1905.

-----. Lamarck and Playfair. In Geological Magazine, 43: 145-53, 193-202. 1906.

Gerstner, Patsy A. A Dynamic Theory of Mountain Building: Henry Darwin Rogers, 1842. In Isis, 66: 26-37. 1975.

Glaessner, M. F. and Teichert, C. Geosynclines: A Fundamental Concept in Geology. In American Journal of Science, 245: 465-82, 571-91. 1947.

Gortani, Michele. Italian Pioneers in Geology and Mineralogy. In Journal of World History, 7: 503-19. 1963.

Graham, R. P. D. The Development of Mineralogical Science. In Proceedings and Transactions of the Royal Society of Canada, Volume 28, Third Series, Section 4: 33-42. 1934.

Greene, John C. and Burke, John G. The Science of Minerals in the Age of Jefferson. In American Philosophical Society Transactions, 68 (4): 1-113. 1978.

Greene, Mott T. Geology in the Nineteenth Century; Changing Views of a Changing World. Ithaca, NY; Cornell University Press. 1982.

Gregory, Herbert E. A Century of Geology -- Steps of Progress in the Interpretation of Land Forms. In American Journal of Science, 196: 104-32. 1918.

Hall, D. H. History of the Earth Sciences During the Scientific and Industrial Revolutions with Special Emphasis on the Physical Geosciences. New York; Elsevier. 1976.

Hallam, Anthony. A Revolution in the Earth Sciences from Continental Drift to Plate Tectonics. Oxford; Clarendon Press. 1973.

Hanna, G. Dallas. Early Reference to the Theory That Diatoms Are the Source of Bituminous Substances. In American Association of Petroleum Geologists. Bulletin, 12 (5): 555-56. 1928.

Hansen, Bert. Early History of Glacial Theory in British Geology. In Journal of Glaciology, 9 (55): 135-41. 1970.

Harrington, John W. The First, First Principles of Geology. In American Journal of Science, 265: 449-61. 1967.

-----. The PreNatal Roots of Geology; a Study in the History of Ideas. In American Journal of Science, 267 (5): 592-7. 1969.

Hausen, Hans. The History of Geology and Mineralogy in Finland 1828-1918. Helsinki; Societas Scientiarum Fennica. 1968.

Hazen, Robert M. The Founding of Geology in America: 1771-1818. Geological Society of America Bulletin, 85 (12): 1827-34. 1974.

-----. North American Geology, Early Writings. Stroudsburg, PA; Dowden, Hutchinson & Ross. 1979.

Hedberg, Hollis D. Influence of Torbern Bergman (1735-1784) on Stratigraphy. In Stockholm Contributions in Geology, 20: 19-47. 1969a.

-----. The Influence of Torbern Bergman (1735-1784) on Stratigraphy: a Résumé. In Toward a History of Geology, Proceedings of the New Hampshire Inter-Disciplinary Conference on the History of Geology, September 7-12, 1967, edited by Cecil J. Schneer. Cambridge, MA; M.I.T. Press. 1969. pp. 186-92.

Higgins, G. E. and Saunders, J.B. Mud Volcanoes -- Their Nature and Origin. In Naturforschende Gesellschaft Basel. Verhandlungen, 84 (1): 101-52. 1974.

Hobbs, William H. Evolution and the Outlook of Seismic Geology. In American Philosophical Society. Proceedings, 48: 259-302. 1909.

Hooykaas, Reijer. Catastrophism in Geology, Its Scientific Character in Relation to Actualism and Uniformitarianism. Koninklijke Nederlandse Akademie van Wetenschappen, afdeling Letterkunde, Mededelingen, 33 (7): 271-316. 1970.

Hubbert, M. King. Critique of the Principle of Uniformity. In Philosophy of Geohistory: 1785-1970, edited by Claude C. Albritton Jr. Stroudsburg, PA; Dowden, Hutchinson and Ross, Inc. 1975. pp 225-55.

Hunt, Charles B. G. K. Gilbert, on Laccoliths and Intrusive Structures. In The Scientific Ideas of G. K. Gilbert, edited by Ellis L. Yochelson. G.S.A. Special Paper 183: 25-34. 1980.

-----. G. K. Gilbert's Lake Bonneville Studies. In The Scientific Ideas of G. K. Gilbert, edited by Ellis L. Yochelson. G.S.A. Special Paper 183: 45-59. 1980.

Hurlbut, Cornelius S. and Switzer, George S. Gemology. New York; Wiley. 1979.

Jahn, Melvin E. Some Notes on Dr. Scheuchzer and on Homo diluvii testis. In Toward a History of Geology, Proceedings of the New Hampshire Inter-Disciplinary Conference on the History of Geology, September 7-12, 1967, edited by Cecil J. Schneer. Cambridge, MA; M.I.T. Press. 1969. pp. 193-213.

Jones, F. Wood. John Hunter as a Geologist. In Annals of the Royal College of Surgeons, 12: 219-44. 1953.

Jordan, William M. Geology and the Industrial-Transportation Revolution in Early to Mid Nineteenth Century Pennsylvania. In Two Hundred Years of Geology in America, Proceedings of the New Hampshire Bicentennial Conference on the History of Geology, edited by Cecil J. Schneer. Hanover, NH; University Press of New England. 1979. pp. 91-103.

Judd, John W. Henry Clifton Sorby and the Birth of Microscopical Petrology. In Geological Magazine, 5: 193-204. 1908.

Karunakaran, C. and Murthy, S. R. N. On Some Earth Science Observations in Sanskrit Texts. In Records of the Geological Survey of India, 109 (2): 82-103. 1977.

Kelly, Sister Suzanne. Theories of the Earth in Renaissance Cosmologies. In Toward a History of Geology, Proceedings of the New Hampshire Inter-Disciplinary Conference on the History of Geology, September 7-12, 1967, edited by Cecil J. Schneer. Cambridge, MA; M.I.T. Press. 1969. pp. 214-25.

Krynine, Paul D. On the Antiquity of Sedimentation and Hydrology. In Geological Society of America Bulletin, 71: 1721-6. 1960.

Kunz, George F. Life and Work of Haüy. In American Mineralogist, 3: 61-89. 1918.

LaRocque, Aurèle. Milestones in the History of Geology; a Chronologic List of Important Events in the Development of Geology. In Journal of Geological Education, 22 (5): 195-203. 1974.

LeConte, Joseph. A Century of Geology. In Annual Report of the Smithsonian Institution, 1900: 265-87. 1901 (reprint).

Lindroth, S. Urban Hiärne 1641-1724. In Swedish Men of Science: 42-9. Stockholm; Swedish Institute. 1952.

Lintner, Stephen F. and Darwin H. Stapleton. Geological Theory and Practice in the Career of Benjamin Henry Latrobe. In Two Hundred Years of Geology in America, Proceedings of the New Hampshire Bicentennial Conference on the History of Geology, edited by Cecil J. Schneer. Hanover, NH; University Press of New England. 1979. pp. 107-119.

Mabey, Don R. Pioneering Work of G. K. Gilbert on Gravity and Isostasy. In The Scientific Ideas of G. K. Gilbert, edited by Ellis L. Yochelson. G.S.A. Special Paper 183: 61-68. 1980.

Manten, A. A. Half a Century of Modern Palynology. In Earth-Science Reviews, 2: 277-316. 1966.

Margerie, Emanuel de. Three Stages in the Evolution of Alpine Geology. In Quarterly Journal of the Geological Society of London, 102: xcvii-cxiv. 1946.

Marsh, O. C. History and Methods of Palaeontological Discovery. In American Journal of Science, Series 3, 18: 323-59. 1879.

Mather, Kirtley and Mason, Shirley. A Source Book in Geology. New York; McGraw Hill. 1939.

Matthews, Edward B. Progress in Structural Geology. In Johns Hopkins University Studies in Geology, 8: 137-61. 1927.

Merrill, George P. Contributions to the History of American Geology. Washington, D.C.; Government Printing Office. 1906.

-----. The First One Hundred Years of American Geology. New Haven, CT; Yale University Press. 1924.

Meyerhoff, A. A. Arthur Holmes: Originator of Spreading Ocean Floor Hypothesis. In Journal of Geophysical Research, 73 (20): 6563-5. 1968.

Miller, Benjamin Leroy. Progress in Ore Genesis Studies. In Johns Hopkins University Studies in Geology, 8: 121-35. 1927.

Miller, Samuel. Brief Retrospect of the Eighteenth Century. New York; T. & J. Swords. 1803.

Morrell, J.B. Reflections on the History of Scottish Science. In History of Science, 12: 81-94. 1974.

Morton, W. L. Henry Youle Hind (1823-1908). In Geological Association of Canada Proceedings, 23: 25-29. 1971.

Murthy, S. R. N. Earth Science Studies in Ancient India. In Indian Minerals, 30: 29-42. 1976.

Murty, K. S. History of Geoscience Information in India. In Geoscience Information: A State of the Art Review, edited by A. P. Harvey and J. A. Diment. London; Broad Oak Press. 1979. pp. 51-60.

Neill, Patrick. Biographical Account of Mr. Williams, the Mineralogist. In Annals of Philosophy, 4: 81-3. 1814.

Neve, Michael and Porter, R. S. Alexander Catcott: Glory and Geology. In British Journal for the History of Science, 10: 47-70. 1877.

North, F. J. Centenary of the Glacial Theory. In Proceedings of the Geological Association, 54: 1-28. 1943.

Ohern, D. W. Fifty Years of Petroleum Geology. In Johns Hopkins University Studies in Geology, 8: 83-120. 1927.

Oldroyd, D. R. Historicism and the Rise of Historical Geology, Parts I and II. In History of Science, 17 (3): 191-213 and 17 (4): 227-257. 1979.

-----. Mineralogy and the 'Chemical Revolution'. In Centaurus, 19: 54-71. 1975.

-----. A Note on the Status of A. F. Cronstedt's Simple Earths and His Analytical Methods. In Isis, 65: 506-12. 1974.

-----. Some NeoPlatonic and Stoic Influences on Mineralogy in the Sixteenth and Seventeenth Centuries. In Ambix, 21 (2-3): 128-56. 1974.

-----. Some Phlogistic Mineralogical Schemes Illustrative of the Concept of 'Earth' in the Seventeenth and Eighteenth Centuries. In Annals of Science, 30: 269-306. 1974.

Ospovat, Alexander M. Reflections on A. G. Werner's "Kurze Klassifikation". In Toward a History of Geology, Proceedings of the New Hampshire Inter-Disciplinary Conference on the History of Geology, September 7-12, 1967, edited by Cecil J. Schneer. Cambridge, MA; M.I.T. Press. 1969. pp. 242-56.

-----. Werner´s Concept of the Basement Complex. In History of Concepts in Precambrian Geology, edited by W. O. Kupsch and W. A. S. Sarjeant. Geological Survey of Canada Special Paper, 19: 161-70. 1979.

Owen, Edgar W. Remarks on the History of American Petroleum Geology. In Journal of the Washington Academy of Science, 49 (7): 256-60. 1959.

-----. Trek of the Oil Finders; a History of Exploration for Petroleum. Tulsa, OK; American Association of Petroleum Geologists. 1975.

Page, Leroy E. Diluvialism and Its Critics in Great Britain in the Early Nineteenth Century. In Toward a History of Geology, Proceedings of the New Hampshire Inter-Disciplinary Conference on the History of Geology, September 7-12, 1967, edited by Cecil J. Schneer. Cambridge, MA; M.I.T. Press. 1969. pp. 257-71.

Phillips, Frank C. An Introduction to Crystallography. Third Edition. New York; Wiley. 1963.

Pirsson, Louis V. Rise of Petrology as a Science. In American Journal of Science, 46: 222-39. 1918.

Porter, Roy S. George Hoggart Toulmin´s Theory of Man and the Earth in the Light of the Development of British Geology. In Annals of Science, 35 (4): 339-52. 1978.

-----. The Making of Geology; Earth Science in Britain, 1660-1815. Cambridge; Cambridge University Press. 1977.

-----. William Hobbs of Weymouth and His "The Earth Generated and Anatomized". In Society for the Bibliography of Natural History Journal, 7: 333-41. 1976.

Press, Frank and Raymond Sever. Earth. San Francisco, CA; W. H. Freeman. 1974.

Prior, G. T. Review of Progress of Mineralogy from 1864 to 1918. In Geological Magazine, 56 (1): 10-18. 1919.

Pyne, Stephen J. Certain Allied Problems in Mechanics: Grove Karl Gilbert at the Henry Mountains. In Two Hundred Years of Geology in America, Proceedings of the New Hampshire Bicentennial Conference on the History of Geology, edited by Cecil J. Schneer. Hanover, NH; University Press of New England. 1979. pp. 225-28.

-----. From the Grand Canyon to the Marianas Trench: the Earth Sciences After Darwin. In The Sciences in the American Context: New Perspectives, edited by Nathan Reingold. Washington, D.C.; Smithsonian Institution Press. 1979. pp. 165-192.

Ramsay, Andrew C. Passages in the History of Geology: Being an Inaugural Lecture, University College, London, 1848. London; Printed for Taylor and Walton. 1848.

-----. Passages in the History of Geology: Being an Introductory Lecture at University College, London, in Continuation of the Inaugural Lecture of 1848. London; Printed for Taylor, Walton and Maberly. 1849.

Ranalli, Georgio. Robert Hooke and the Huttonian Theory. In Journal of Geology, 90 (3): 319-23. 1982.

Regnell, Gerhard. On the Position of Palaeontology and Historical Geology in Sweden. In Arkiv för Mineralogi och Geologi, 1 (1): 1-64. 1947.

-----. Primaeval Earth as Conceived by 18th Century Naturalists. In History of Concepts in Precambrian Geology, edited by W. O. Kupsch and W.A.S. Sarjeant, Geological Association of Canada Special Paper 19: 171-80. 1979.

Rodolico, Francesco. Niels Stensen, Founder of the Geology of Tuscany. In Dissertations on Steno as Geologist. In Acta Historica Scientiarum Naturalium et Medicinalium, 23: 237-43. 1971.

Rowlinson, J. S. Theory of Glaciers. In Royal Society of London Notes and Record, 26 (2): 189-204. 1971.

Rudwick, Martin J. S. The Meaning of Fossils. Second Edition. New York; Neale Watson. 1976.

-----. The Strategy of Lyell's Principles of Geology. In Isis, 61 (1): 5-33. 1970.

Rupke, Nicolaas A. Continental Drift Before 1900. In Nature, 227: 349-50. 1970.

Sarjeant, William A. S. and Harvey, A. P. Uriconian and Langmyndian: A History of the Study of the Precambrian Rocks of the Welsh Borderland. In History of Concepts in Precambrian Geology, edited by W. O. Kupsch and W. A. S. Sarjeant. Geological Association of Canada Special Paper, 19: 181-224. 1979.

Sarton, George. Introduction to the History of Science. Volumes 1-3. Baltimore, MD: Published for the Carnegie Institution of Washington by Williams and Wilkins Co. 1927.

Scherz, Gustav. Niels Stensens Riesen. In Dissertations on Steno as Geologist. In Acta Historica Scientiarum Naturalium et Medicinalium, 23: 9-137. 1971.

Schneer, Cecil J. Introduction. In Toward a History of Geology, Proceedings of the New Hampshire Inter-Disciplinary Conference on the History of Geology, September 7-12, 1967; edited by Cecil J. Schneer. Cambridge, MA; M.I.T. Press. 1969. pp. 1-18.

-----. Rise of Historical Geology. In Isis, 45: 256-67. 1954.

Schuchert, Charles. A Century of Geology; the Progress of Historical Geology in North America. In American Journal of Science, 196: 45-103. 1918.

Seddon, George. Abraham Gottlob Werner; History and Folk-history. In Geological Society of Australia Journal, 20 (4): 381-95. 1973.

Segonzac, G. Dunoyerde. Birth and Development of the Concept of Diagenesis (1866-1966). In Earth Science Reviews, 4: 153-201. 1968.

Shastri, Vaidyanath. Sciences in the Vedas. New Delhi; Sarvadeshik Arya Pratinidhi Sabha. 1970.

Siegfried, Robert and R. H. Dott, Jr. Humphry Davy as Geologist, 1805-29. In British Journal for the History of Science, 9: 219-27. 1976.

Simpson, George G. Uniformitarianism; an Inquiry into Principle, Theory and Method in Geohistory and Biohistory. In Philosophy of Geohistory: 1785-1970, edited by Claude C. Albritton, Jr. Stroudsburg, PA: Dowden, Hutchinson and Ross, Inc. 1975.

Skinner, Herbert C. Raymond Thomassy and the Practical Geology of Louisiana. In Two Hundred Years of Geology in America, Proceedings of the New Hampshire Bicentennial Conference on the History of Geology, edited by Cecil J. Schneer. Hanover, NH; University Press of New England. 1979. pp. 201-11.

Sleep, Mark C. W. Sir William Hamilton (1730-1803), His Work and Influence in Geology. In Annals of Science, 25: 319-38. 1969.

Stoddart, D. R. Darwin, Lyell, and the Geological Significance of Coral Reefs. In British Journal for the History of Science, 9 (2): 199-218. 1976.

Stokes, Evelyn. The Six Days and The Deluge; Some Ideas on Earth History in the Royal Society of London 1660-1775. In Earth Science Journal, 3 (1): 13-39. 1969.

Stroh, A. H., editor. Emmanuel Swedenborg as a Scientist. Stockholm; Aftonbladets Trycheri. 1908-11.

Sweet, Jessie M. and Waterston, C. Robert Jameson´s Approach to the Wernerian Theory of the Earth, 1796. In Annals of Science, 23: 81-95. 1967.

Taylor, John. Report on the State of Knowledge Respecting Mineral Veins. In Report of the British Association, Third Meeting, 1833: 1-25. 1834.

Taylor, Kenneth L. Nicolas Desmarest and Geology in the Eighteenth Century. In Toward a History of Geology, Proceedings of the New Hampshire Inter-Disciplinary Conference on the History of Geology, September 7-12, 1967, edited by Cecil J. Schneer. Cambridge, MA; M.I.T. Press. 1969. pp.339-56.

Thams, Johann C., ed. Development of Geodesy and Geophysics in Switzerland. Zurich; Swiss Academy of Natural Science. 1967.

Thomas, H. Hamshaw. Rise of Geology and Its Influence on Contemporary Thought. In Annals of Science, 5: 325-41. 1947.

Thwaites, Fredrik T. Development of the Theory of Multiple Glaciation in North America. In Transactions of the Wisconsin Academy of Science, Arts and Letters, 23: 41-164. 1927.

Tikhomirov, V. V. Development of Geological Sciences During the First Half of the Nineteenth Century. In International Geological Congress, Report of the Twenty-third Session, Czechoslovakia, Section 13. Proceedings: 319-29. 1968.

-----. The Development of the Geological Sciences in the U. S. S. R. from Ancient Times to the Middle of the Nineteenth Century. In Toward a History of Geology, Proceedings of the New Hampshire Inter-Disciplinary Conference on the History of Geology, September 7-12, 1967, edited by Cecil J. Schneer. Cambridge, MA; M.I.T. Press. 1969. pp. 357-85.

Tomkieff, Sergei I. Classification of Igneous Rocks. In Geological Magazine, 76: 41-8. 1939.

-----. A Historical Survey of Petrology. 1936 edition translated by Frants Levinson-Lessing. Edinburgh; Oliver and Boyd. 1954.

-----. Unconformity -- An Historical Study. In Geologists´ Association, London. Proceedings. 73 (4): 383-417. 1962.

Wallace, Robert E. G. K. Gilbert´s Studies of Faults, Scarps, and Earthquakes. In The Scientific Ideas of G. K. Gilbert, edited by Ellis L. Yochelson. G. S. A. Special Paper 183: 35-44. 1980.

Waterston, Charles D. William Thomson (1761-1806) a Forgotten Benefactor. In University of Edinburgh Journal, 22 (2): 122-34. 1965.

Wegmann, Eugene. Changing Ideas about Moving Shorelines. In Toward a History of Geology, Proceedings of the New Hampshire Inter-Disciplinary Conference on the History of Geology, September 7-12, 1967, edited by Cecil J. Schneer. Cambridge, MA; M.I.T. Press. 1969. pp. 386-414.

White, George W. Early Description and Explanation of Kettle Holes; Charles Whittlesey (1808-1886). In Journal of Glaciology, 5 (37): 119-22. 1964.

-----. John Josselyn´s Geological Observations. In Illinois Academy of Science Transactions, 48: 173-82. 1954.

-----. Lewis Evans´ Contributions to Early American Geology, 1743-1755. In Illinois Academy of Science Transactions, 44: 152-58. 1951.

-----. Lewis Evans´ Early American Notice of Isostasy. In Science, 114 (2960): 302-3. 1951.

-----. William McClure´s Concept of Primitive Rocks ("Basement Concept"). In History of Concepts in Precambrian Geology, edited by W. O. Kupsch and W. A. S. Sarjeant. Geological Association of Canada Special Paper, 19: 251-9. 1979.

Wilson, John A. Darwinian Natural Selection and Vertebrate Paleontology. In Symposium on the History of American Geology, December 27 and 28, 1958. In Journal of the Washington Academy of Sciences, 49 (7): 231-33. 1959.

Wood, Robert M. Coming Apart at the Seams (Alfred Wegener, the Founder of Continental Drift -- Or Was He?). In New Scientist, 85 (1191): 252-54. 1980.

Woodward, Horace B. History of Geology. New York; G. P. Putnam's Sons. 1911.

Yanshin, A. L., Garetskiy, R. G., and Shiezinger, A. Y. Role of the U. S. S. R. Academy of Sciences in Development of the Theory of Platforms, and Some Present Aspects of This Theory. In Geotectonics, 3: 127-36. 1974.

Yochelson, Ellis L. Charles D. Walcott -- America's Pioneer in Precambrian Paleontology and Stratigraphy. In History of Concepts in Precambrian Geology, edited by W. O. Kupsch and W. A. S. Sarjeant. Geological Association of Canada Special Paper, 19: 261-92. 1979.

Yoder, Hatten S. Jr. Experimental Mineralogy: Achievements and Prospects. In Bulletin de Minéralogie, 103 (1): 5-26. 1980.

Zittel, Karl A. von. History of Geology and Palaeontology to the End of the Nineteenth Century. London; Walter Scott. 1901.

AUTHOR INDEX

Dawson, John Wilson
1864

Day, Jeremiah
1810

Delafosse, Gabriel
1840

Delesse, Achille Ernest
Oscar Joseph
1847, 1857-58

Delius, Christoph Teaugott
1770

Democritus of Abdera
Fourth/Fifth Century BC

Derham, William
1713

Descartes, René
1633, 1637, 1644, 1647

Deshayes, Gerard-Paul
1831

Desmarest, Nicholas
1760, 1762, 1765, 1794, 1836

Desor, Edouard
1852

Dobrzensky, Jacobus Johnnes Wenceslaus
1657

Dobson, Peter
1825

Dolomieu, (Deodat) Gratet de
1781, 1791

Douglas, James
1785

Dufrenoy, Ours Pierre
Armand Petit
1844

Duhamel, Jean Baptiste
1660

Duns, Johannes Scotus
Thirteenth Century

Dupiut, Jules-Juvenal
1863

Durocher, Joseph Marie Elisab
1841, 1845-46, 1857

Dutton, Clarence Edward
1876, 1880, 1882, 1889, 1890

Eaton, Amos
1818, 1830

Ebel, Johann Gottfried
1809

Egen, Peter Nikolaus Casper
1828

Ehrenberg, Christian Gottfrie
1838, 1839

Emmons, Ebenezer
1826, 1842-43, 1852,
1854, 1858

Emmons, Samuel Franklin
1886

Empedocles of Agrigentum
Fifth Century BC

Epicurus
Fourth Century

Eschscholtz, Johann Friederic
1821

Esmark, Jens
1824

Eustatius
c. 440

Evans, Evan Wilhelm
1864, 1866

Evans, Lewis
1743

Falloppius, Gabriel
1557, 1564

Farey, John
1809